Jingshi Li

Optical Delay Interferometers and their Application for Self-coherent Detection

Karlsruhe Series in Photonics & Communications, Vol. 11
Edited by Profs. J. Leuthold, W. Freude and C. Koos

Karlsruhe Institute of Technology (KIT)
Institute of Photonics and Quantum Electronics (IPQ)
Germany

Optical Delay Interferometers and their Application for Self-coherent Detection

by
Jingshi Li

Dissertation, Karlsruher Institut für Technologie (KIT)
Fakultät für Elektrotechnik und Informationstechnik, 2012

Impressum

Karlsruher Institut für Technologie (KIT)
KIT Scientific Publishing
Straße am Forum 2
D-76131 Karlsruhe
www.ksp.kit.edu

KIT – Universität des Landes Baden-Württemberg und
nationales Forschungszentrum in der Helmholtz-Gemeinschaft

KIT Scientific Publishing 2013
Print on Demand

ISSN 1865-1100
ISBN 978-3-86644-957-2

Optical Delay Interferometers and their Application for Self-coherent Detection

Zur Erlangung des akademischen Grades eines

DOKTOR-INGENIEURS

der Fakultät für Elektrotechnik und Informationstechnik
des Karlsruher Instituts für Technologie

genehmigte

DISSERTATION

von

M.Sc. Jingshi Li

aus

Liaoning, V. R. China

Tag der mündlichen Prüfung: 12.11.2012
Hauptreferent: Prof. Dr. sc. nat. Juerg Leuthold
Korreferenten: Prof. Dr.-Ing. Dr. h. c. Wolfgang Freude
Prof. Dr. rer. nat. Uli Lemmer

Table of Contents

List of Figures

List of Tables

Abstract (German)

Selbst-kohärente Empfänger sind vielversprechende Kandidaten für Empfänger in optischen Netzwerken mit 100 Gbit/s. Diese Empfänger bestehe aus einer Mehrzahl von Delay-Interferometern (DI) mit breitbandigen Photodioden an ihren Ausgängen. Schnelle echtzeit digitale Signalprozessierung (DSP) ermöglicht den kohärenten Empfang von optischen Signalen, unter denen das polarisationsgemultiplexte (PolMUX) differenzielle Quadrature Phase Shift Keying (DQPSK) als am besten geeignet für 100 Gbit/s-Netzwerke erachtet wird.

Verglichen mit einem konventionellen kohärenten Empfänger, benötigt ein selbst-kohärenter keinen Lokaloszillator. Diese sind kostspielig, da sie geringe Linienbreite und hohe Frequenzstabilität aufweisen müssen. Dagegen sind selbst-kohärente Empfänger relativ komplex und ihre Symbolrate muss für gewöhnlich im Vorfeld durch die Länge der Delay-Interferometer festgelegt werden. Außerdem werden meist weitere optische Bauelemente, z.B. ein Polarisationsverfolger, benötigt um PolMUX-Signale zu empfangen.

In dieser Arbeit fokusieren wir uns auf die Entwicklung eines Delay-Interferometers und dessen Anwendung in der selbst-kohärenten Empfangstechnik. Zunächst wird ein abstimmbares Delay-Interferometer mit geringer Polarisationsbahängigkeit eingeführt. Dieses erlaubt den selbst-kohärenten Empfang bei verschiedenen Symbolraten. Anschließend wird eine kompakte Umsetzung eines selbst-kohärenten Empfängers vorgestellt: Vier verschachtelte DIs formen das optische frontend (OFE) für diesen polarisations und phasensenstiven selbst-kohärenten Empfänger. Diese Konfiguration reduziert die Komplexität des Empfängers signifikant und vereinfacht ebenso dessen Kontrolle. Zusätzlich entwickeln wir einen DSP-Algorithmus zum selbst-kohärenten Empfang und demultiplexen von Signalen auf zwei orthogonalen Polarisationen. Zum Schluss wird ein Gesamtkonzept aus OFE und DSP-Algorithmus für selbst-kohärenten Empfang von fortgeschrittenen Singalformaten, insbesondere 100 Gbit/s PolMUX DQPSK, präsentiert.

Summary

Self-coherent receivers are a promising candidate for data reception in 100 Gbit/s optical networks. Self-coherent receivers consist of multiple delay interferometers (DI) with high-speed photodiodes at the outputs. By applying digital signal processing (DSP) techniques it is possible to detect coherent optical signal formats, among which the polarization multiplexed (PolMUX) differential quadrature phase shift keying (DQPSK) technique is considered to be the format of choice for 100 Gbit/s networks.

Compared to the conventional coherent receiver self-coherent receivers do not require a local oscillator. Local oscillators are expensive as they need to have a narrow line-width and a high frequency stability. Conversely, self-coherent receivers have a relatively complex configuration and their operating symbol-rate is usually fixed by the built-in delay interferometers (DI). Further, to detect a PolMUX signal, additional optical devices, e.g., a polarization tracker are usually required for polarization demultiplexing.

In this thesis, we focus on the development of delay interferometers and their application to self-coherent reception. Firstly, a tunable delay interferometer with low polarization dependence is introduced to allow for self-coherent at various symbol rates. Then a compact interferometer is presented. It comprises four folded DIs and two folded optical hybrids. It works as an optical frontend (OFE) for both self-coherent and coherent reception with polarization and phase diversity. This configuration significantly reduces the complexity of the receiver and simplifies its control. In addition, we develop a DSP algorithm for self-coherent reception to demultiplex signals at two orthogonal polarizations. We finally present a complete self-coherent scheme together with OFE and DSP algorithms for the reception of advanced signal formats, and in particular for detecting 100 Gbit/s PolMUX DQPSK signals.

The thesis is organized as follows. In Chapter 1, the state-of-the art and the motivation for this thesis is presented. Firstly a brief review on coherent and self-coherent receiver techniques is given. Then the challenges and problems with self-coherent receivers are listed, and the solutions are discussed in the rest of the thesis.

In Chapter 2, we firstly give a brief introduction on digital modulation formats and multiplexing techniques as used in optical communications. Then we explain the principle of direct detection, differential direct detection, coherent and self-coherent receiver techniques. To analyze the polarization dependence of our free-space DI, we discuss the polarization rotation of an incident beam on different optical interfaces in Section 2.3. With this knowledge, we discuss the optical elements used in the DI system, Section 2.4.

In Chapter 3, we present the operating principle of a tunable DI whose polarization dependent frequency shift (PDFS) is mitigated by means of birefringent elements. We further explain the method for accurately controlling the DI in terms of time delay and phase offset between its two arms. This part of the thesis has been published in Optics Express [J5] and several conference proceedings [C6][C13][C19][C21]. In the second half of Chapter 3, we introduce a compact interferometer configuration, where 4 DIs used in a polarization and phase diverse self-coherent receiver are folded into one DI structure. The configuration can also work as a coherent receiver optical frontend by adjusting the built-in optical waveplates. A minimization of the receiver OFE is done by using a micro-optical bench fabricated with LIGA (X-ray lithography, electroplating (galvanic), and molding (Abformung)). This part of the thesis has been filed as a patent [P1] and submitted to the Optics Express [J1].

Chapter 4 presents the digital signal processing algorithms for self-coherent reception. Firstly, the accuracy of conventional field recovery algorithms is tested with tunable DIs. For demultiplexing the PolMUX signal we improve the field and polarization recovery algorithm in two steps. In the first step, the field recovery algorithm is combined with an equalizer. By using the filtered signal as a corrector, the signals in two polarizations can be demultiplexed if polarizations are not coupled too strongly. To overcome this limitation, in the second step, a decision feedback is used to remove the accumulated noise in the recovered field and a training sequence is introduced to estimate the polarization change caused by the channel. A variant of a PolMUX DQPSK signal comprising a normal DQPSK and a 45° offset DQPSK is introduced to demonstrate reception of a 100 Gbit/s PolMUX DQPSK signal for arbitrary polarization couplings. The content of this chapter has been published in two conference proceedings [C6][C21] and an Optics Express paper [J2].

In Chapter 5 we show that besides differential direct detection and self-coherent detection, tunable delay interferometers can also be used for many other applications in optical communications. They may be used for performing the optical fast Fourier transform (FFT) at ultra-fast speed [J4][J7][C7][C11][C12][C14][C16][C17]. They can be used for all-optical wavelength conversion to reshape pulses [J9][C22] [C24][C25][74], or they may be used for dispersion compensation [83].

In the Appendix, we firstly review the "constant modulus" and the so called "decision-direct least-mean square" equalizer algorithms for demultiplexing signals transmitted on two polarizations of a PolMUX signal, A.1. The the theory of the interferometer frontend is presented, A.2. The signal processing of a phase diverse delay interferometer pair is mathematically explained in A.3. The training sequence for channel estimation is given in A.4. At the end, transmitter configurations for PolMUX signal comprising normal DQPSK and 45° offset DQPSK signal are presented in A.5.

Achievements of the Present Work

In this thesis, state-of-the-art delay interferometers and self-coherent receivers have been investigated. In the following, we give a concise overview of the main achievements.

Tunable Delay Interferometer: A free-space optical delay interferometer (DI) has been designed and built. It is capable of tuning the time delay between two arms from 0 ps to 100 ps, and adjusting the phase offset between the two arms with a step size of <1°. This allows detection of differential phase shift keying signals at various symbol rates [C19][J5]. The DIs can also be used for self-coherent reception of signals at different symbol rates and at different sampling rates [C6][C13][C21]. It can be utilized in many other applications in optical communions including optical FFT for ultra-fast optical signal processing [J4][J7][C7][C11][C12][C14][C16][C17], and for all optical wavelength conversion [J9][C22] [C24][C25][74].

PDFS Mitigation in Delay Interferometer: The polarization rotation for optical interfaces is theoretically analyzed. To compensate the resulting polarization-dependent phase shift, a birefringent element is inserted in one optical arm of the delay interferometer. This way, a polarization-dependent frequency shift (PDFS) of the DI's transfer function can be minimized. This allows the delay interferometer operation in a differential direct detection receiver or in a phase-diverse self-coherent receiver with input signals at arbitrary states of polarization [C13][C19][C21][J5].

Control of Tunable Delay Interferometer: A independent optical pilot tone is introduced to accurately control and lock the time delay and phase offset of the DI [C13][J5].

Polarization and Phase-Diverse Self-coherent Receiver: We reduce the complexity of setup and control by employing free-space micro optics, which folds four DIs into a simple structure with only one active control. The design was published in [P1][J1][J5][C6]. This configuration can also be switched into a polarization and phase diverse coherent receiver with addition of a local laser oscillator. A miniaturized design is achieved by using a LIGA micro-optical bench.

Field Reconstruction with Self-coherent Receiver: A decision feedback loop is added to a recursive-division field reconstruction algorithm [J2][J3]. The field can be reconstructed with high accuracy without using any additional amplitude detection channel. The noise accumulation is removed from the recovered field [J2] after the decision circuit.

Polarization Demultiplexing with Self-coherent Receiver: Two solutions have been discussed to demultiplex the signals carried by two orthogonal polarizations. In the first solution, a butterfly equalizer is positioned after the conventional field reconstruction algorithm to perform channel estimation and polarization demultiplexing (PolDEMUX). Then the filtered signal is used to correct the recovered field [C6]. This method can demultiplex a PolMUX signal without optical polarization tracker. However, it fails when the signals in two polarizations are extremely mixed. To further improve the performance, in the second solution, training sequences for channel estimation are investigated for self-coherent reception. Then the butterfly equalizer is combined with the decision feedback to separate the signals of two orthogonal polarizations [J2]. For the first time, a self-coherent system that can detect signals at arbitrary states of polarization is demonstrated with a 112 Gbit/s PolMUX DQPSK signal.

1 Introduction

New modulation techniques together with new reception techniques are under intense investigation because of an increasing capacity demand for optical channels in the global network. In the last five years, the global IP traffic increased eightfold to 30,734 PB per month. It is expected to increase threefold from 2011 to 2016 with a compound annual growth rate of 29 % [1]. A large amount of data traffic originates from video related applications. Especially, mobile network traffic has been increasing with a growth rate of 133 % in 2011. Mobile traffic is estimated to overtake wired network traffic and is likely becoming the leading drive of the data traffic increase [2]. In order to deal with traffic demands, optical networks with single carrier bit rate of 100 Gbit/s have been standardized by IEEE 802.3ba. In a dense wavelength division multiplexing (DWDM) network, this improves the transmission capacity of a single fiber to 8.8 Tbit/s. To further increase the line capacity, 400 Gbit/s single carrier transceivers are now under discussion. They are capable to provide a line rate of > 23 Tbit/s [3]. Moreover super-channel using optical orthogonal division multiplexing (OFDM) technique has been introduced to push the line capacity to even higher bit rates [J7].

Optical signals are traditionally modulated with an "on-off keying" (OOK) format. They encode the signal by blocking and unblocking the beam. However, this is not the optimal solution for 100 Gbit/s or 400 Gbit/s networks anymore, because they require a much narrower pulse width which pushes both the transmitter and receiver electronics to the limit. Therefore advanced signal formats which modulate not only the optical amplitude, but also the phase, frequency and polarization state of the signal are of the interest to increase the spectral efficiency of the signal. Among the advanced modulation formats, quadrature phase shift keying (QPSK) and differential quadrature phase shift keying (DQPSK) with two bits per symbol have been shown as promising candidates not only because of their increased spectral efficiency but also due to the good transmission properties like improved polarization mode dispersion (PMD) tolerance, relaxed optical signal to noise ratio (OSNR) requirements and better robustness against nonlinearities than binary OOK [4]. To further increase the spectral efficiency, polarization multiplexing (PolMUX) is a common technique to encode data on the two orthogonal polarization states of the signal [5]. A short introduction of advanced modulation formats and multiplexing technique will be given in Chapter 2.1.

Optical signal intensity detection is traditionally done by using a single photodiode. For advanced signal formats, as information is also encoded in the phase and the polarization state of the optical carrier, new reception technologies are required. The most popular reception methods include differential direct detection, coherent detection and self-coherent detection.

The principle of differential direct detection will be presented in more detail in Chapter 2.2. Here we briefly summarize the state of the art on differential direct detection. Differential direct detection comprising delay interferometers (DI) has been introduced in the first place because of the simple structure. It combines the signal with itself in interferometers where the signal and a copy delayed by one symbol are compared with certain phase offset. The information encoded in the differential phase between signal and copy is then converted into intensity signals which are then fed into decision circuits. In 2002, Gnauck showed a differential binary phase shift keying (DBPSK) transmission experiment with single delay interferometer [6], and in the same year Griffin presented a DQPSK transmission experiment [4] with two DIs. Differential direct detection is, however, limited to only detecting signals with constant modulus at a fixed bit rate. Because one DI is only able to distinguish two differential phase states, the number of DIs used for differential direct detection also increases with the number of phase states in the modulation format. These disadvantages make differential direct detection impractical for receiving high level modulation formats. Besides' this, differential direct detection also requires an optical polarization tracker to detect a PolMUX signal [5][7]-[10]. A high-speed endless and accurate polarization tracker is costly, and it increases the footprint of the receiver. Other methods not requiring an optical polarization tracker have been investigated. In [11], the signals on two polarizations are firstly interleaved in time and then detected at twice the symbol rate. Though no optical polarization tracker is needed, this method requires twice the bandwidth of the photo detectors and electronics in the receiver. In [12], a variant of a DI is introduced. In addition to differential direct detection of the signals on two orthogonal polarizations, the detection scheme optically cross-couples the two polarizations so that the cross product between the two orthogonally polarized signals can be formed by digital signal processing (DSP). Compared to the conventional differential direct detection scheme, this method has the disadvantage that it requires twice as many photodiodes and analog-to-digital converters (ADC).

Coherent receivers for detection of coherent signal were studied intensively during the 1980s [13]-[23]. They superimpose the signal with a laser which serves as a local oscillator (LO). The theory of coherent detection will be discussed in Chapter 2.2. Here we only briefly review the state of the art. Compared to differential direct detection, coherent detection has the advantage of an increased receiver sensitivity of 2.3 dB [24]. Another advantage is that it provides a frequency shift of the optical spectrum which allows for an easy signal processing in the electrical domain. Phase and polarization diverse coherent reception was firstly proposed in 1987 [25]. Recently this scheme has been reinvestigated and demonstrated for PolMUX QPSK of signals up to 100 Gbit/s [26][27] and also demonstrated for other advanced formats, e.g., 512QAM (quadrature amplitude modulation) [28] and 1024QAM [29]. Compensation of signal impairments and polarization demultiplexing for PolMUX signals can be achieved with advanced DSP algorithms [30]. The main limitation of coherent reception is that it requires a high quality laser serving as a local oscillator. Furthermore, phase and frequency offsets between the signal and LO need to be compensated.

The concept of a self-coherent receiver was presented by the group of Kikuchi [31] and X. Liu [32] in 2006. Compared to coherent receivers, a local oscillating laser is no longer required. Self-coherent receivers mix the signal with a delayed copy of itself in a phase diverse delay interferometer pair. Differing from the conventional differential direct detection method, the electric field of the signal is reconstructed from the outputs of the delay interferometers digitally. Advanced DSP algorithms yield a reception sensitivity improvement as well as chromatic dispersion compensation [33]. Field recovery can be divided into phase recovery and amplitude recovery. The phase recovery is relatively easy. To derive the phase, the accumulated phase offsets from the differential phases over time need to be summed, starting with an arbitrary point in time. Amplitude recovery may be achieved in different ways. The amplitude can be measured with an external amplitude branch, such as performed by Kikuchi [31]. The advantage is that it allows self-coherent reception for detecting multilevel signals, e.g., 16QAM [34], the disadvantage is that additional photodiodes and additional electronics, e.g., ADC, are required. Solution not requiring additional amplitude branch was proposed by X. Liu [32]. The amplitude is estimated by using a geometric mean of the adjacent fields at the DI outputs. However, it is only suitable for signal formats with constant modulus, e.g., DBPSK, DQPSK, and D8PSK.

The challenges in developing a self-coherent receiver for the reception of advanced modulation formats such as 100 Gbit/s PolMUX DQPSK signals are in both its optical frontend and the digital signal processing (DSP) detection algorithms. More precisely, they are as follows:

- Tunable time delay of delay interferometers: In order to operate at various symbol rates and sampling rates, self-coherent reception requires a tunable time delay within the built-in DIs. A continuously tunable DI has been proposed in [35], where the delay is introduced by passing two orthogonal polarizations through a tunable birefringent element. This scheme, however, requires the input signal to have equal power for two orthogonal polarizations. In this thesis we choose a solution based on free-space optics as optical path adjustment is most easily done in free space. Details will be described in Chapter 3.1.

- Reduction of polarization sensitivity of the delay interferometer: Similar as differential direct detection, a phase diverse self-coherent receiver needs to be polarization independent. Delay interferometers with low polarization dependence and high stability are therefore of interest. Prior to building a polarization insensitive DI we first perform a theoretical analysis on the polarization rotation of an optical interface in Chapter 2.3. Further, the analysis of the optical elements used in the delay interferometer will be given in Chapter 2.4. Since birefringence is hard to avoid despite all efforts, so we introduced additional birefringent elements to compensate for the residual polarization dependent frequency shift (PDFS). This is discussed in Chapter 3.1. Measurement results with self-coherent receiver algorithms will be shown in Chapter 4.1.

- Complexity reduction of a polarization and phase diverse receiver: Polarization and phase diverse self-coherent receivers in principle consist of 4 separate DIs which all need to be controlled accurately. A reduction of the structure complexity and control is necessary for commercial systems. One method is to use two polarization insensitive DIs and only then separate the polarizations at the outputs [33]. However, polarization-insensitive DIs are intricate to fabricate. A second method consists in combining the I (inphase) and Q (quadrature phase) DIs. This technique works well with planar lightwave circuit (PLC) technology, e.g., with 2×4 MMIs (multimode interference coupler) [36][37]. While the PLC technology is well suited for the mass market, it is inflexible in that the time delay in integrated optics cannot be continuously tuned. Conversely, solutions based on free-space optics offer good thermal stability, low insertion loss, a large operation wavelength range and true delay tunability. Yet, combining I and Q demodulation in one DI is not straightforward. In Chapter 3.2, a micro-optical interferometer will be introduced to fold 4 DIs of the I and Q channels for two orthogonal polarizations in one single DI structure where a single piezo-motor is used to adjust the time delay and phase shift of 4 DIs. We will further show that by adjusting the built-in waveplates, this configuration can be easily switched into a polarization and phase diverse coherent receiver.

- Amplitude recovery without using an external amplitude branch: Amplitude recovery need to be performed with high accuracy in order to detect signal formats with varying amplitude, e.g., a quadrature amplitude modulation (QAM) signal or a PolMUX signal. In [J3], we introduced a recursive division method which in principle can recover the amplitude of the signal without using an additional amplitude branch. However, the method suffers from noise accumulation. A training sequence was then introduced to reset the recursive division algorithm when the noise exceeds a threshold [J3]. A 16QAM signal reception is simulated with this method. However, this comes at the cost of extra redundancy because of the training sequence. In Chapter 4.3, we will introduce a decision feedback loop which is combined with the recursive division circuit to remove the accumulated noise during the field recovery.

- Polarization demultiplexing without optical polarization tracker: Similar as the differential direct detection, self-coherent receivers conventionally require optical polarization trackers to receive a PolMUX signal. In Chapter 4.2, we will present a polarization recovery algorithm. By using a butterfly equalizer, the polarization change caused by the channel is compensated, and the filtered signal is then used as a corrector to reduce the accumulated noise. However, this method does not work well when the polarizations are too strongly coupled. To further improve the algorithm, a training sequence will be introduced in Chapter 4.3 to better estimate the channel. The recursive division method combined with a decision feedback loop will be used to improve the field recovery accuracy. Another challenge is that the zero points in the field lead to an infinite output during the recovery process. In Chapter 4.3, we will discuss this issue in detail and introduce a variant of a PolMUX DQPSK comprising a normal DQPSK and a 45° offset DQPSK. At the end, an experimental demonstration of the 112 Gbit/s PolMUX DQPSK signal reception will be presented.

2 Fundamentals

In this chapter, we will firstly have a brief introduction on the digital modulation formats and multiplexing techniques in optical communications. Then we will give an overview on existing receiver techniques. As the delay interferometer discussed in this thesis is based on free-space optics, in the third part of this chapter, we will present a theoretical analysis on polarization rotation on different optical interfaces and then we will give a brief summary of the optical elements in the delay interferometer.

2.1 Digital Modulation Formats and Multiplexing Techniques

The electric field of an optical signal $\vec{E}(t)$ can be expressed with components in two orthogonal polarization states x and y,

$$\vec{E}(t) = \begin{bmatrix} E_x(t) \\ E_y(t) \end{bmatrix} = \begin{bmatrix} A_{0x}(t)\cos\left(\omega_{0x}(t)t + \varphi_{0x}(t)\right) \\ A_{0y}(t)\cos\left(\omega_{0y}(t)t + \varphi_{0y}(t)\right) \end{bmatrix} = \mathrm{Re}\left\{ \begin{bmatrix} A_{0x}(t)e^{j\left(\omega_{0x}(t)t + \varphi_{0x}(t)\right)} \\ A_{0y}(t)e^{j\left(\omega_{0y}(t)t + \varphi_{0y}(t)\right)} \end{bmatrix} \right\}. \qquad (2.1.1)$$

where $E_x(t)$ and $E_y(t)$ are the electric field components of two orthogonally polarized lights. The amplitude of $E_x(t)$ and $E_y(t)$ are denoted with $A_{0x}(t)$ and $A_{0y}(t)$, and the phase $\varphi_{0x}(t)$ and $\varphi_{0y}(t)$. The optical carrier frequency is $\omega_{0x,y}(t)$. In the phasor notation, the optical signal is denoted with its carrier frequency $\omega_{0x,y}(t)$ and complex amplitude $A_{x,y}(t) = A_{0x,y}(t)\exp\left[j\varphi_{0x,y}(t)\right]$. An optical signal thus has four degrees of freedom: amplitude, phase, frequency and polarization. Besides these, there is another degree of freedom in spatial domain, which is, however, out of the discussion in this thesis. One can encode information on any or multiple of these variables. This leads to digital modulation formats such as amplitude-shift keying (ASK), phase-shift keying (PSK), frequency-shift keying (FSK), and polarization shift keying (PolSK), whereas a combination of ASK and PSK is known as quadrature amplitude modulation (QAM). Among them, the ASK on-off keying (OOK) modulation format is most common in commercial telecommunication systems because of its simplicity. Recently, two differential PSK (DPSK) formats, namely differential binary phase-shift keying (DBPSK) and differential quadrature phase-shift keying (DQPSK) have been considered as a replacement of OOK because of their better tolerance against chromatic dispersion (CD) and fiber nonlinearities [4][6]. In addition, DQPSK increases the spectral efficiency (SE) of the signal by a factor of two. Multi-level signals, e.g., 16QAM, are of special interest in light of further increasing the SE.

Due to CD limitation, fiber nonlinearities and also the finite bandwidth of electrical devices, optical single carrier signals cannot fully utilize the capacity of the optical fiber. The solution is to multiplex several signal-channels onto one fiber. Multiplexing techniques include wavelength-division multiplexing (WDM), and polarization multiplexing (PolMUX).

Among them, WDM is a common technology standardized by industry. WDM also allows for PolMUX so that two orthogonal polarization states can be encoded and thus double the SE. To reduce the channel spacing and further increase SE, subcarrier multiplexing technology, i.e., orthogonal frequency division multiplexing (OFDM) has been discussed recently [38]. Other multiplexing techniques including space-division multiplexing [39] among which multi-core fiber systems have emerged only recently, however, are not in the scope of this thesis.

In summary, in order to increase the capacity of an optical transmission link one has to combine different technologies, i.e., combine multi-level modulation formats encoding with multiplexing techniques.

In this section, a brief review on OOK, PSK and QAM modulation formats as well as multiplexing technologies such as WDM, OFDM and PolMUX will be given.

2.1.1 On-off Keying Signal

On-off keying signal is the simplest form of ASK signal formats. Dating back to ancient times, optical signal had already been widely used for communications. In Fig. 2.1, the left figure shows beacons on the Great Wall in China and the right figure presents a border tower that belonged to the Roman Empire nowadays located in Germany. The beacons and border tower were used to alert the defense troops of an enemy by lighting an optical source. In another word, they have two states. One is 'on', which carries the information that the enemy is nearby, the other is 'off' that there is no enemy around. This might be the earliest form of on-off keying signal (OOK).

Fig. 2.1. Left: beacons on the Great Wall in China; Right: border tower used in the Roman Empire located in Germany (Photos are obtained from internets).

In the modern optical communications, OOK was and still is widely used globally. Instead of manually controlling with a beacon, an optical-electrical transmitter driven by electrical signal is used for signal modulation. The transmitter may be a direct modulated laser or a continuous wave (CW) laser connected with a modulator, e.g., based on a Mach-Zehnder interferometer (MZI). Among which the external modulation with MZI is preferred because it can be operated at high speed. In the MZI, the phase of an optical signal is modified by the applied electrical time dependent signal that causes either a constructive or destructive interference at the output of MZI. The constructive interference gives a signal with "on" state and the destructive interference "off" state.

In Fig. 2.2, an example of OOK signal is illustrated. On the left, it depicts the signal over time. At different symbol durations (T_s) the signal is denoted with "1" when it's "on" state, and with "0" when it's "off" state. On the right side of the figure, a constellation diagram of the complex amplitude of the OOK signal is depicted.

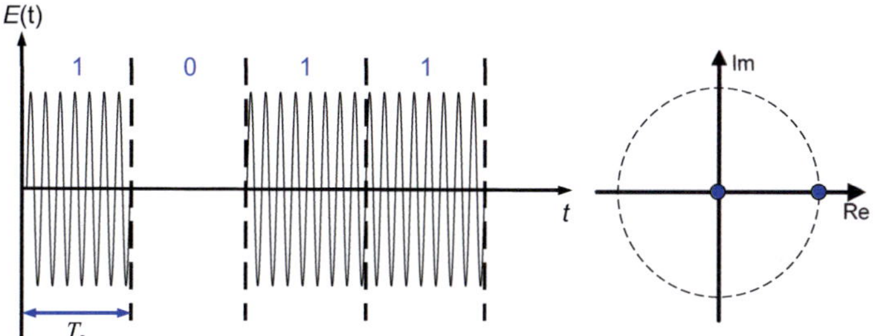

Fig. 2.2. On-off keying signal. Left: signal over time; Right: constellation diagram.

After transmission via fiber, the signal is detected by a single photodiode where the optical signal is converted into the electrical domain. The fundamental principle of the signal detection is optical absorption. More details can be found in Section 2.2.1.

The signal can come with a non-return to zero (NRZ) or return-to-zero (RZ) pulse shape. The difference between the two pulse shapes is shown in Fig. 2.3. Instead of occupying the whole time slot such as with an NRZ signal, the "on" state of the RZ signal only occupies part of the symbol slot and return to zero within each symbol.

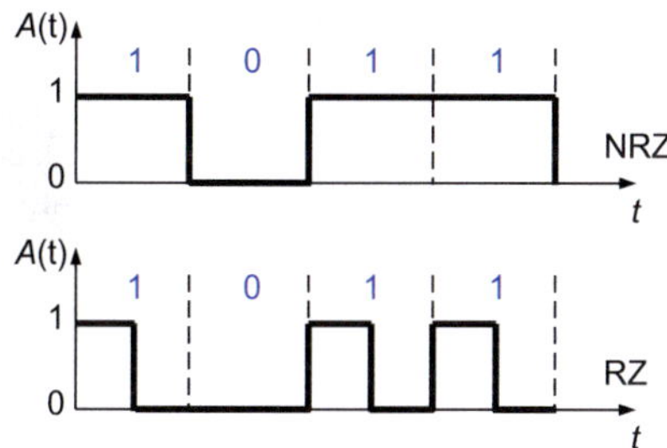

Fig. 2.3. The upper figure depicts a NRZ-OOK signal over time; The lower figure depicts a RZ-OOK signal over time.

NRZ format has the advantage that its bandwidth is smaller than RZ format by about a factor of 2 simply because on-off transitions occur fewer times. However, RZ format has a receiver sensitivity advantage of about 1-3 dB due to the higher peak power and the better tolerance against inter-symbol interference (ISI) [40].

2.1.2 Phase Modulation Formats

Phase shift keying (PSK) signals are generated by modulating the phase of an optical signal, while the amplitude is kept constant. NRZ-DBPSK signal is generated by modulating the optical phase with two phase states "0" and "π", while the intensity ideally remains constant. For a DBPSK signal (see Fig. 2.4(a)), as the phase is encoded differentially, a phase change indicates a "1" and no phase change a "0". For a BPSK signal, phase state "0" indicates a "0" and phase state "π" a "1". To increase the spectral efficiency, (D)QPSK signal sends two bits per symbol using four phase states or differential phases (0, $\pi/2$, π, $-\pi/2$), Fig. 2.4(b). A further increase of SE can be achieved by signal formats like D8PSK, Fig. 2.4(c).

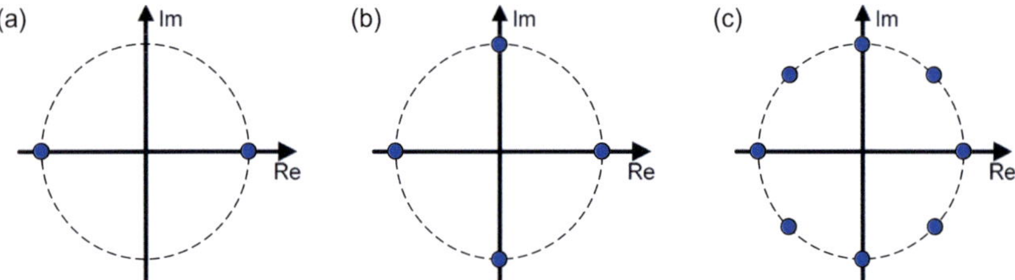

Fig. 2.4. Constellation diagrams of PSK/DPSK signals, (a) (D)BPSK, (b) (D)QPSK, and (c) (D)8PSK.

The PSK signal can be generated with a phase modulator, or an IQ-modulator with MZIs operating with push-pull mode, or a combination of both.

The reception of DPSK signal can be simply done by differential direct detection using delay interferometers [4][6]. It can also be detected with coherent [26][27] and self-coherent [31][32][33] receiver. More details for the receiver techniques are in Section 2.2.

2.1.3 Quadrature Amplitude Modulation

Quadrature amplitude modulation (QAM) is a modulation scheme in which the binary data are represented by different amplitudes and phases. The following figures give some examples of different QAM signals.

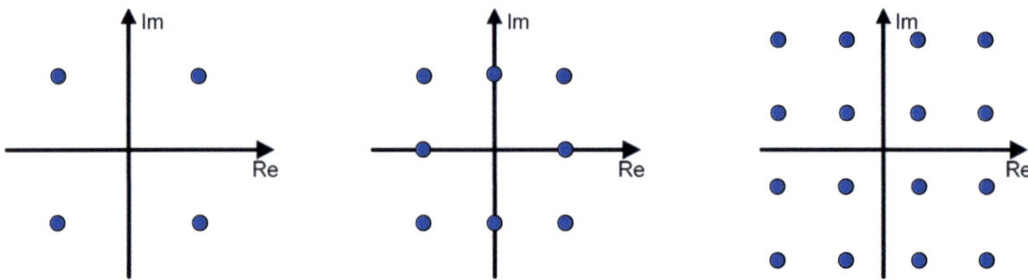

Fig. 2.5. Constellation diagrams of QAM signals, (a) 4QAM, (b) 8QAM, and (c) 16QAM.

An M-ary QAM signal can carry more than one bit in a time-period. Every single point in the four quadrants represents a symbol carrying $\log_2(M)$ bits. They have different amplitude levels and different phases. Compared with OOK signals, quadrature amplitude modulation signals have higher spectral efficiency.

The generation of QAM can be achieved by a combination of phase modulator and MZI [41], or an IQ-MZI modulator which requires multi-level electrical driving signals [42].

The reception of QAM is conventionally done by a coherent receiver. Recently, reception with self-coherent receiver is also reported [34][J3].

2.1.4 Wavelength Division Multiplexing

Wavelength division multiplexing (WDM) combines independent optical signals at different optical carrier wavelengths into the same optical fiber. There are two low loss transmission windows in an optical fiber. As shown in Fig. 2.6, one is around 1300 nm and the other is around 1550 nm (C and L bands). The two windows have in total a bandwidth of 27 THz. Compared to single carrier signal, WDM signal can utilize the channel capacity with a significant higher efficiency.

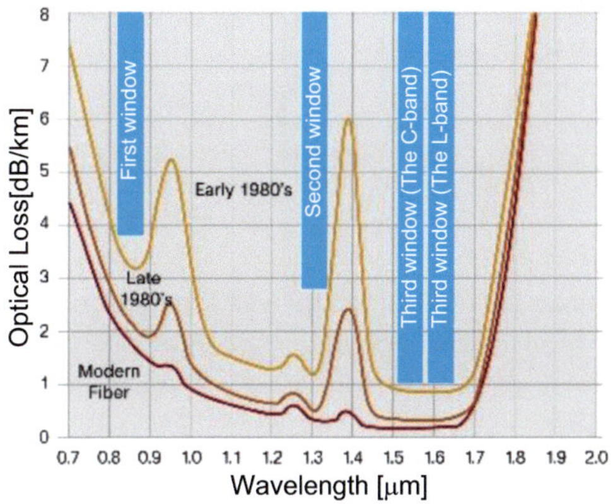

Fig. 2.6. Low–loss transmission windows of silica fibers in the wavelength regions near 1300 nm and 1550 nm [43].

The WDM signal is generated by combining independent optical carriers with optical multiplexers and detected with demultiplexers. According to ITU (International Telecommunication Union) standards, WDM systems can be classified into coarse WDM (CWDM) and dense WDM (DWDM) systems. The channel spacing of CWDM is 20 nm. The channel spacing of DWDM is reduced to 100 GHz, 50 GHz or 12.5 GHz.

2.1.5 Orthogonal Frequency Division Multiplexing

Orthogonal frequency division multiplexing (OFDM) is another kind of multi carrier modulation scheme. The subcarriers of OFDM are generated following the rule that they are orthogonal to each other. Such a modulation scheme has advantage in the spectral efficiency and minimizes the inter-carrier interference [38]. Because the overall symbol rate is low, a relatively long symbol in time with the help of cyclic prefix makes the OFDM signal more tolerant against chromatic dispersion [38].

We write a baseband OFDM signal symbol, which has a symbol duration of T_s, as

$$s(t) = \begin{cases} \mathrm{Re}\left\{ \sum_{i=0}^{N-1} D_i \exp(\mathrm{j}2\pi(f_0 + i\Delta f)t) \right\}, & -\dfrac{T_s}{2} < t < \dfrac{T_s}{2} \\ 0, & \text{else} \end{cases}, \tag{2.1.2}$$

where N is the number of subcarriers, $D_i (i=1, 2, \ldots, N)$ is the data encoded on each subcarrier, T_s is the OFDM symbol duration, f_0 the center frequency of the first subcarrier, and Δf the subcarrier spacing. We can calculate the cross correlation between two subcarriers at f_n and f_m as

$$\begin{aligned} R_{\mathrm{xcorr}} &= \frac{1}{T_s} \int_{-T_s/2}^{T_s/2} e^{\mathrm{j}2\pi f_n t} \left(e^{\mathrm{j}2\pi f_m t} \right)^* dt = \frac{1}{T_s} \int_{-T_s/2}^{T_s/2} e^{\mathrm{j}(2\pi M\Delta f)t} dt \\ &= \frac{1}{\mathrm{j}2\pi M\Delta f T_s} \left(e^{\mathrm{j}(\pi M\Delta f)T_s} - e^{-\mathrm{j}(\pi M\Delta f)T_s} \right) = \frac{\sin\left(\pi M\Delta f T_s \right)}{\pi M\Delta f T_s}, \end{aligned} \tag{2.1.3}$$

where $M\Delta f$ is frequency difference between the two subcarriers, and '*' denotes the for complex conjugate operator. When $M\Delta f T_s$ is a non-zero integer, i.e., $\Delta f = 1/T_s$, $R_{\mathrm{xcorr}} = 0$, the two subcarriers are orthogonal to each other. Thus for OFDM signal, we have the subcarrier center frequency defined as

$$f_i = \frac{i}{T_s} \qquad (i = 0, 1, \ldots, N-1). \tag{2.1.4}$$

In Fig. 2.7(a), an instance of OFDM symbol with four subcarriers without modulation is presented. Each subcarrier has same amplitude and phase. After modulation, depending on the modulation scheme the amplitude and phase of the subcarriers may be different from each other (see Fig. 2.7(b)).

Putting this symbol into frequency domain, we have its real part,

$$S(f) = \sum_{i=0}^{N-1} D_i \frac{\sin\left[\pi T_s (f_0 + i\Delta f - f) \right]}{\pi (f_0 + i\Delta f - f)}. \tag{2.1.5}$$

The sinc function is the result of the rectangular pulse shape. For each subcarrier, its peak locates at the center frequency of the subcarrier, as shown in Fig. 2.8. For each subcarrier, all the other subcarriers are equal to zero at its center frequency. Therefore they do not interfere with each other.

The modulation and demodulation of OFDM signal can be achieved by using inverse discrete Fourier transform (IDFT) and discrete Fourier transform (DFT). For the effective application one can use the inverse fast Fourier transform (IFFT) as well as fast Fourier transform (FFT) to reduce the computational complexity remarkably.

Advanced coding for OFDM signal developed in wireless communications can be directly applied to optical communications to mitigate polarization dependent loss (PDL) in the optical channel [C4][C5].

The OFDM signal can be up-converted into optical domain directly with an IQ-MZI modulator or first up-converted onto an intermediate frequency and then to optical carrier [38]. An all-optical OFDM signal can be generated using a comb laser source [J4], and demodulated with cascaded delay interferometers. Details of the all-optical OFDM are given in Chapter 5.1.

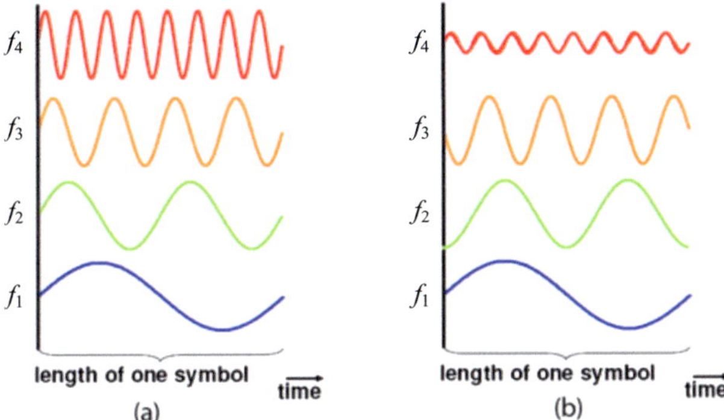

Fig. 2.7. OFDM signal at separate subcarrier in time. (a) Carriers without modulation. (b) Modulated carriers [44].

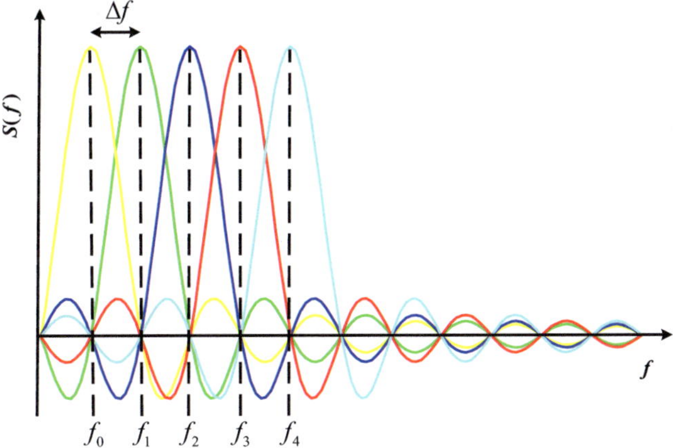

Fig. 2.8. Five subcarriers of an OFDM signal of one symbol in the frequency domain. They have equal carrier spacing Δf.

2.1.6 Polarization Multiplexing

Polarization multiplexing (PolMUX) is a common technique to increase the spectral efficiency of an optical signal. It multiplexes two independent signals at the same WDM frequency band onto two orthogonal polarization states. It can be either bit aligned as shown in Fig. 2.9(a) or bit interleaved in Fig. 2.9(b). For bit interleaved scheme, the signal has to be RZ shaped.

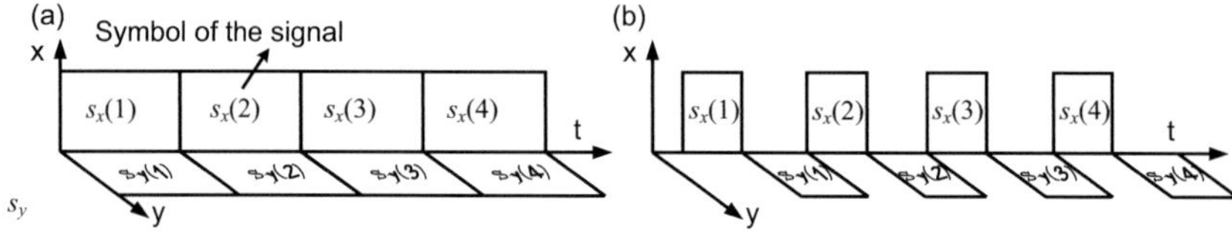

Fig. 2.9. Two kinds of polarization multiplexing scheme: (a) bit aligned polarization multiplexing, (b) bit interleaved polarization multiplexing.

Experimental record of 3.2 bit/s/Hz per wavelength using polarization multiplexed DQPSK has been reported [45][46].

Since the state of polarization of a signal after fiber transmission is uncertain and time-varying, the reception of a polarization multiplexed signal requires adaptive polarization demultiplexing, either through optical means [5][7]–[10], or electronic means [12][30]. A brief introduction of the electronic means is given in Appendix A.1.

2.2 Receiver Techniques

The role of an optical receiver is to convert the optical signal into electrical domain in order to perform further signal processing and demodulation. The optical receiver can be classified into direct detection receivers, differential direct detection receivers, coherent detection receivers and self-coherent detection receivers. In the following sections, the principle of each type of the receiver will be briefly introduced.

2.2.1 Direct Detection Receiver

Optical direct detection is the most commonly used reception technique in fiber-optic communication systems, because it leads to the simplest design for transmitters and receivers. The fundamental mechanism behind direct detection is optical absorption in a photodiode.

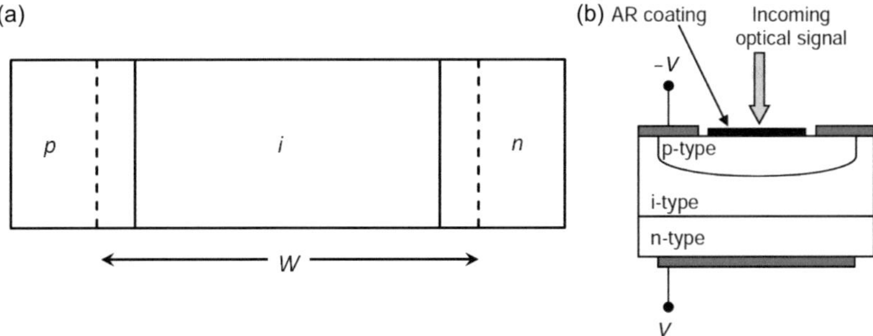

Fig. 2.10. (a) Structure of a p-i-n photodiode; (b) Design of a p-i-n photodiode with anti-reflective (AR) coating [47].

A photodiode is operated in reverse-bias. A reverse-biased p-n junction consists of a large depletion region, which essentially is devoid of free charge carriers and which comprises of a large built-in electric field that opposes the flow of electrons from n (negative) -side to p (positive) -side and of holes from p to n. When the depletion region is illuminated with light, electron-hole pairs are generated through absorption. Because of the built-in electric field, most electrons and holes accelerate in opposite directions and drift to the n- and p-sides without recombination. The resulting flow of current I_p is proportional to the incident optical power P_{in} [47],

$$I_p = \frac{\eta e}{hf} P_{in},$$

(2.2.1)

where η is quantum efficiency. It is defined as:

$$\eta = \frac{\text{electron-hole pair generation rate}}{\text{photon incidence rate}}.$$

(2.2.2)

In order to increase η, an antireflection (AR) coating can be coated to the incident facet of the photodiode (see Fig. 2.10(b)). Another way to increase η is to increase depletion-region width W by inserting an intrinsic region sandwiched between p- and n-type layers, as shown in Fig. 2.10(a).

The incident optical power is related to the signal intensity I_{sig} by,

$$I_{sig} = \frac{P_{in}}{A_{eff}}, \tag{2.2.3}$$

where A_{eff} is the effective area of the beam. If the electric field of a signal is defined as,

$$E(t) = \mathrm{Re}\left[A(t)e^{j\omega_0 t} \right]. \tag{2.2.4}$$

The corresponding intensity is

$$I_{sig} = \langle E \times H \rangle = \sqrt{\frac{\varepsilon}{\mu}} \frac{1}{T} \int_0^T |E(t)|^2 \, dt = \frac{1}{2}\sqrt{\frac{\varepsilon}{\mu}} |A(t)|^2, \tag{2.2.5}$$

where T is a time duration which is much larger than the signal period $2\pi/\omega_0$, ε is the electric permittivity of the medium and μ the magnetic permeability of the medium. $\langle x \rangle$ denotes time averaging of x.

In summary, a direct detector will measure,

$$I_{\mathrm{p}} = \frac{\eta e}{2hf} A_{\mathrm{eff}} \sqrt{\frac{\varepsilon}{\mu}} |A(t)|^2. \tag{2.2.6}$$

Further, the bandwidth of a photodiode in a receiver is of interest as well. It determines the speed with which it responds to variations in the incident optical power. The bandwidth depends on the time taken by electrons and holes to travel to the electrical contacts and also depends on the response time of the electrical circuit used to process the photocurrent.

2.2.2 Differential Direct Detection Receiver

A photodiode can only detect the intensity of an optical signal. In order to convert the phase information into intensity, interferometry has been applied in the optical receiver technology. Differential direct detection is the simplest form of such receiver.

Differential direct detection is usually accompanied with differential encoding. A schematic drawing of a transmitter with a specific differential encoder [48] is depicted in Fig. 2.11.

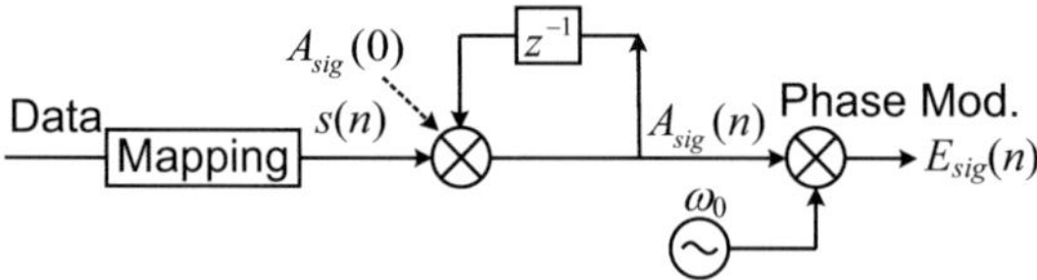

Fig. 2.11. Transmitter with differential encoder. The symbol z^{-1} (representing the z-transform) stands for a time delay by one bit.

The data is firstly mapped to complex symbols $s(n)$ at discrete times $t_n = nT_s$ at multiples n of the symbol period T_s. Then the input symbol sequence $s(n)$ is delayed by one symbol. The delayed copy and original are multiplied generating differentially encoded symbols iteratively $A_{sig}(n) = A_{sig}(n-1)s(n)$. Thus the information is encoded on the change of the electric field.

Then an electrical signal which is proportional with $A_{sig}(n)$ is encoded onto an optical carrier ω_0 by an optical modulator, at the output of the modulator we have an optical signal $E_{sig}(t) = A_{sig}(t)\cos(\omega_0 t)$. Take DBPSK for example, a complete transmitter and receiver system is shown in Fig. 2.12.

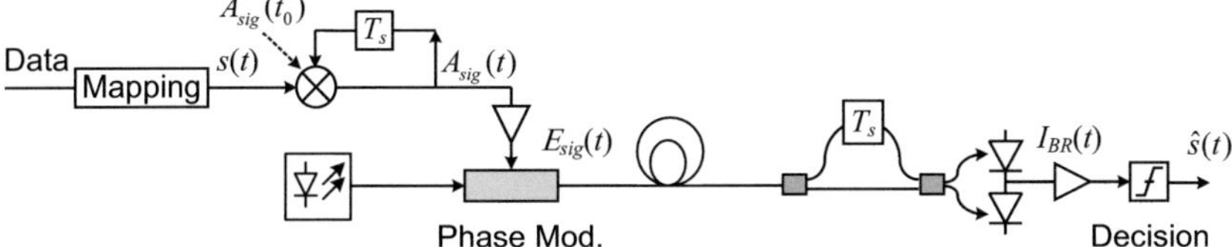

Fig. 2.12. Transmitter and receiver of an optical DBPSK transmission system.

The receiver comprises of an optical DI as its optical frontend. In the DI, the signal is delayed by one symbol duration "T_s" and combined with itself, at the outputs of the DI,

$$I_{out1}(t) \propto \frac{1}{T}\int_0^T \left|E_{sig}(t) + E_{sig}(t - T_s)\right|^2 dt$$

$$\propto \left|A_{sig}(t)\right|^2 + \left|A_{sig}(t - T_s)\right|^2 + 2\left|A_{sig}(t)\right|\left|A_{sig}(t - T_s)\right|\cos\left(\varphi_{0sig}(t) - \varphi_{0sig}(t - T_s)\right),$$

$$I_{out2}(t) \propto \frac{1}{T}\int_0^T \left|E_{sig}(t) - E_{sig}(t - T_s)\right|^2 dt \tag{2.2.7}$$

$$\propto \left|A_{sig}(t)\right|^2 + \left|A_{sig}(t - T_s)\right|^2 - 2\left|A_{sig}(t)\right|\left|A_{sig}(t - T_s)\right|\cos\left(\varphi_{0sig}(t) - \varphi_{0sig}(t - T_s)\right),$$

where $\varphi_{0sig}(t) - \varphi_{0sig}(t - T_s)$ is the phase difference between the two electric fields. Thus, behind the balanced receiver, the differential output current is

$$I_{BR}(t) = I_{out1}(t) - I_{out2}(t) \propto \left|A_{sig}(t)\right|\left|A_{sig}(t - T_s)\right|\cos\left(\varphi_{0sig}(t) - \varphi_{0sig}(t - T_s)\right)$$

$$\propto \mathrm{Re}\left[A_{sig}(t)A_{sig}^{*}(t - T_s)\right] \propto \mathrm{Re}\left[A_{sig}(t - T_s)s(t)A_{sig}^{*}(t - T_s)\right] \tag{2.2.8}$$

$$\propto \left|A_{sig}(t - T_s)\right|^2 \mathrm{Re}\left[s(t)\right].$$

As for DBPSK, at the symbol center, ideally $\left|A_{sig}(t - T_s)\right| = 1$, and $s(t)$ is real, thus the output current of the balanced receiver is proportional to the symbol $s(t)$ at the transmitter. It is then amplified and sent to a decision circuit. Finally an estimated signal $\hat{s}(t)$ is received.

The decision can be illustrated by a line of decision threshold in a constellation diagram as shown in Fig. 2.13(a). Assuming the phase states have same noise level, we draw a line along the imaginary axis (Im). Any point on the right side of the line would result in a positive I_{BR} denoted as '0' (when the differential phase is '0'). Any point on the left side would result in a negative I_{BR} denoted as '1' (when the differential phase is 'π').

To detect DPSK signals with more than two phase states, it requires more than one delay interferometers at the receiver. For example, for DQPSK reception, two DIs are needed, see Fig. 2.13(b). The phase offsets of the DIs are set to be '$-\pi/4$' and '$\pi/4$'. After the two DIs, there are two decision circuits. The decision lines in the constellation diagram are rotated from the imaginary axis by an angle which is equal to the phase offset in DI. In this way two bits of each symbol can be demodulated. For reception of D8PSK signal, four DIs are needed [49]. The phase offsets are set to be '$-3\pi/8$', '$\pi/8$', '$-\pi/8$', and '$3\pi/8$'. One symbol consists of three bits, among which two bits can be directly demodulated after the DIs while one bit requires an XOR (exclusive or) gate in addition, see Fig. 2.13(c).

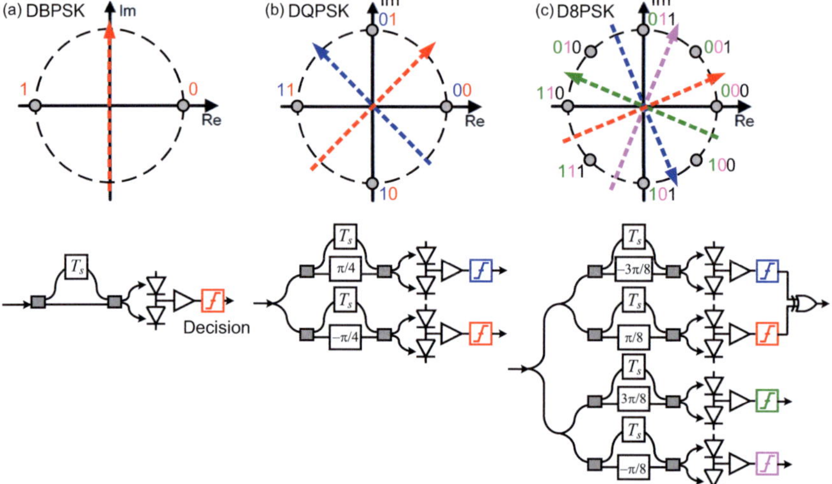

Fig. **2.13**. (a) DBPSK signal constellation diagram and its receiver scheme, (b) DQPSK signal constellation diagram and receiver scheme, (c) D8PSK signal constellation diagram and receiver scheme.

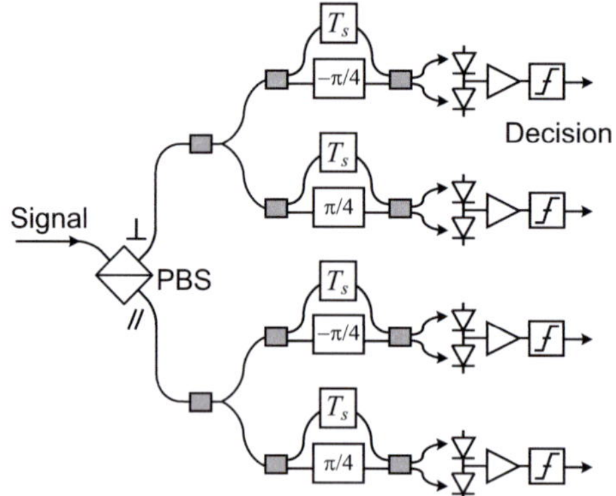

Fig. **2.14**. Polarization diverse differential direct detection receiver for PolMUX-DQPSK signal.

Delay interferometers could have various imperfections, such as low extinction ratio (ER), high PDL and PDFS, which degrades the performance of the receiver [50]. Especially PDFS prevents the receiver working with a signal at arbitrary polarization state. This will be solved in Section 3.1.3 [C19][J5]. A polarization diverse differential direct detection receiver does not require an optical frontend with low PDL or PDFS. In Fig. 2.14, a polarization diverse differential direct detection receiver for PolMUX-DQPSK signal is depicted.

Because the state of polarization can experience random change via the channel link, the signal needs to be firstly aligned to its original polarization state with a polarization controller. Then a polarization beam splitter (PBS) is positioned to split the input signal onto two orthogonal polarizations. The signal at each polarization goes into two DIs with phase offsets '$-\pi/4$' and '$\pi/4$' and then are sent to decision circuits.

A high-speed endless and accurate polarization tracker is costly and it increases the footprint of the receiver. Other methods not requiring an optical polarization tracker have been investigated. In [11], the signals on two polarizations are firstly interleaved in time and then detected at twice the symbol rate. Though no optical polarization tracker is needed, this method requires twice the bandwidth of the receiver photo detectors and electronics. In [12], a variant DI is introduced. It not only differentially direct detects the signals on two orthogonal polarizations, but it also cross-couples the two polarizations optically so that the cross product between the two orthogonally polarized signals can be formed by digital signal processing. Compared to the conventional differential direct detection scheme, this method has the disadvantage that it requires twice as many photodiodes and ADCs.

In summary, with the increase of the phase states of the DPSK signal, the differential direct detection receiver requires more DIs. This makes the receiver structure complicated and impractical for fabrication and control. To receive a PolMUX signal, the differential direct detection receiver either requires a high accurate endless polarization tracker or more complicated optical and electronic setup. To overcome these two limitations, coherent detection receiver and self-coherent detection receiver are introduced.

2.2.3 Coherent Detection Receiver

Coherent detection allows the recovery of the full electric field but requires the implementation of a local oscillator as a reference.

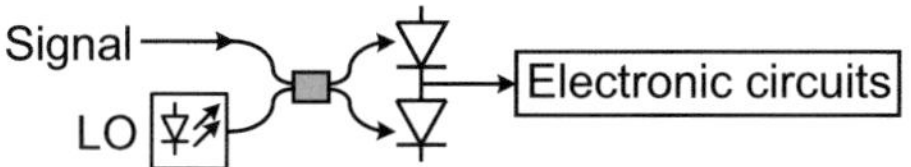

Fig. 2.15. Schematic illustration of a coherent detection receiver.

As shown in the above figure, the input signal and the local oscillator are superimposed in a coupler and detected by an optical detector where interference is taking place. Here the polarization states of the two signals are assumed to be the same. Similar as in Eq. (2.2.7) and Eq. (2.2.8), however, instead of the delayed signal, a local oscillator $E_{LO} = A_{LO}\cos(\omega_{LO}t)$ interferes with the signal, thus at the output of the balanced receiver,

$$I_{\mathrm{BR}}(t) \propto \left|A_{sig}(t)\right|\left|A_{LO}(t)\right|\cos\left[(\omega_0 - \omega_{LO})t + (\varphi_{0sig}(t) - \varphi_{0LO}(t))\right]. \qquad (2.2.9)$$

The signal amplitude is "amplified" by the LO so that the receiver sensitivity is better than with differential direct detection. The phase of the detected signal is related to the phase difference $(\varphi_{0sig}(t) - \varphi_{0LO}(t))$ between the signal and LO, and their carrier frequency difference $(\omega_0 - \omega_{LO})$. This frequency difference is called intermediate frequency (IF).

Depending on the value of IF, a coherent detection receiver is classified as *homodyne*, *intradyne* and *heterodyne* receiver.

As shown in the Table 2.1, when IF = 0, in other words, if the LO have the same frequency as the optical carrier of the signal, the receiver is called a homodyne receiver. In practice, it requires a wavelength tunable laser to be adjusted to the signal frequency, and an optical phase locked loop (PLL) to compensate the time varying phase difference between the two carriers. This makes it impractical to be deployed in commercial systems. In the case of heterodyne and intradyne receivers the frequency of the LO has an offset with respect to the signal carrier. This relaxes the requirement on the local oscillator laser. However, it requires electronic circuits to compensate the frequency offset and phase drift [51]–[53]. The difference between intradyne and heterodyne receivers is the magnitude of IF. Compared to the heterodyne receiver, IF of an intradyne receiver is smaller than the bandwidth of the signal 'B'. As a result, it requires a lower bandwidth of the electronic circuit to compensate the frequency offset and phase shift. This makes intradyne coherent receiver attractive.

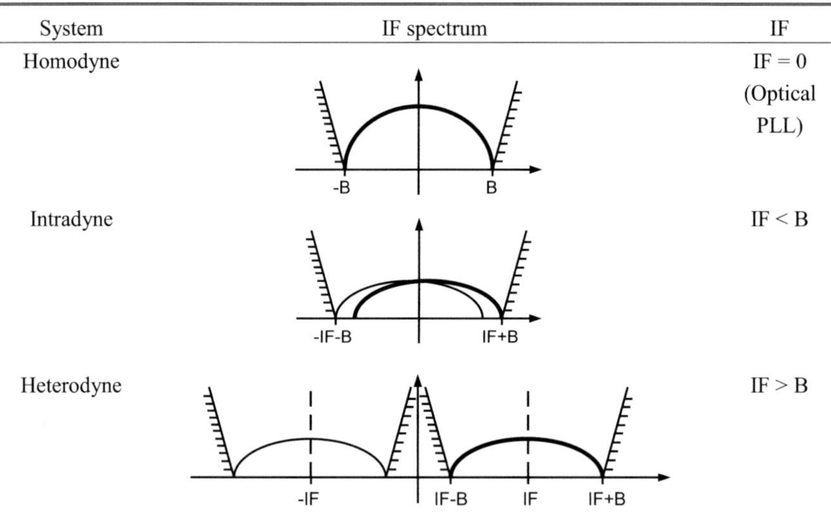

System	IF spectrum	IF
Homodyne		IF = 0 (Optical PLL)
Intradyne		IF < B
Heterodyne		IF > B

Table 2.1. Coherent optical transmission systems (IF = Intermediate frequency, B = Bandwidth of baseband signal) [51].

To fully recover the complex amplitude of the input signal, a phase diverse intradyne receiver is introduced in Fig. 2.16(a). The signal is split into two paths. One is mixed with the local oscillator resulting an inphase (I) signal while the other is mixed with the local oscillator with a $\pi/2$ phase shift resulting a quadrature phase (Q) signal. The principle can be further explained with an example of optical implementation as depicted in Fig. 2.16(b).

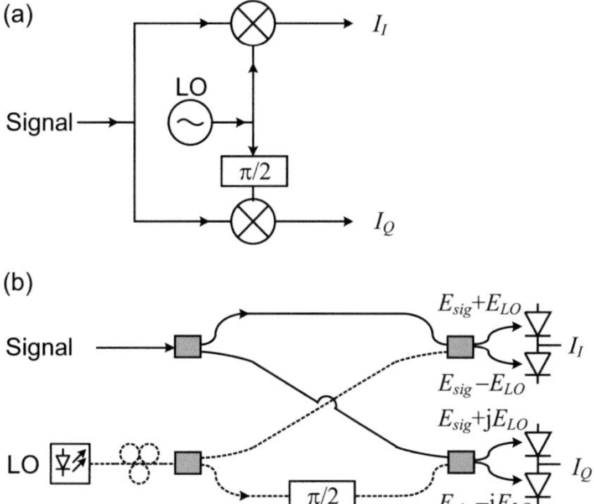

Fig. 2.16. Phase diverse intradyne coherent detection receiver. (a) Schematic illustration, (b) one example of optical implementation.

The signal (E_{sig}) and the local oscillator (E_{LO}) are mixed in a 90 degree optical hybrid. The polarization state of the local oscillator needs to be aligned with the input signal. For convenience, from now on we always denote the signal and the local oscillator with their complex notation,

$$E_{sig} = A_{sig}e^{j\omega_0 t}, \quad E_{LO} = A_{LO}e^{j\omega_{LO} t}. \tag{2.2.10}$$

The currents at the balanced outputs can be derived to

$$I_I(t) \propto \left|E_{sig}(t)\right|\left|E_{LO}(t)\right|\cos\left(\angle E_{sig}(t) - \angle E_{LO}(t)\right),$$
$$I_Q(t) \propto \left|E_{sig}(t)\right|\left|E_{LO}(t)\right|\sin\left(\angle E_{sig}(t) - \angle E_{LO}(t)\right), \tag{2.2.11}$$

where $\angle x$ denotes the phase of a complex signal x. The input signal is demodulated into an inphase and a quadrature phase component, which are orthogonal to each other. The orthogonality can be seen in Eq. (2.2.11) with the sine and cosine functions. We can also write them as a complex signal,

$$u(t) = I_I(t) + jI_Q(t) = E_{sig}(t)E_{LO}^*(t) = \left|A_{sig}\right|\left|A_{LO}\right|e^{j\left[(\omega_0 - \omega_{LO})t + (\varphi_{0sig}(t) - \varphi_{0LO}(t))\right]}, \tag{2.2.12}$$

where '*' denotes a complex conjugate operator. Thus, the amplitude and phase of the received signal can be extracted from the photo currents $I_I(t)$ and $I_Q(t)$.

A natural extension is a polarization diverse receiver as shown in Fig. 2.17. Input signal and local oscillator are first split in two orthogonal polarizations by polarization beam splitters (PBS) respectively. One polarization is perpendicular to the plane formed by the incident direction and the normal to the splitting layer and is called s light ($\perp$). The other polarization is in parallel with the plane, and is p light ($//$). Then the signals at two polarizations are sent into two optical 90° hybrids where the input signal and the local oscillator at each polarization are superimposed. At the outputs, signals of the inphase and quadrature phase components at each polarization are detected. As the signal at the input experienced random change of polarization state during transmission in the optical fiber link, a polarization demultiplexing technique is required at the receiver. Instead of using an optical polarization tracker, recently, with the development of fast electronic circuits, DSP algorithms that are widely used in wireless communications for compensation channel crosstalk of a MIMO (multiple-input and multiple-output) system, have been introduced into optical communications to perform the polarization demultiplexing digitally [30].

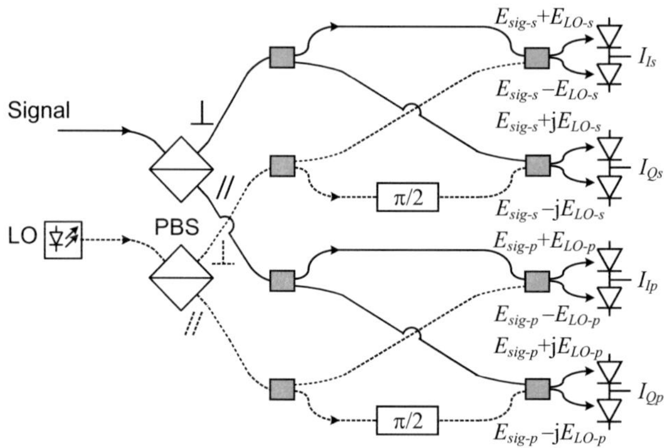

Fig. 2.17. Schematic illustration of a polarization and phase diverse intradyne coherent detection receiver. PBS: polarization beam splitter. $\perp$: s light; $//$: p light.

2.2.4 Self–coherent Detection Receiver

Self-coherent detection receiver was firstly presented by a group at Hitachi [31] and a group at Bell labs [32] in 2006. Instead of using a local oscillator laser in the coherent detection receiver, in self-coherent detection receiver the signal interferes with its delayed copy. Compared to conventional differential direct detection receivers, self-coherent detection receivers utilize advanced DSP algorithms in electronic circuits for signal processing and demodulation, which allows a significant reduction of complexity in the optical frontend for the receiver. In this case, even for DPSK / PSK signals with more than two phase states, only two DIs are needed.

A schematic drawing of a self-coherent detection receiver is depicted in Fig. 2.18. Firstly, the incoming signal is equally distributed onto two paths and then each of them goes into an optical DI with delay equal to τ. The two DIs are arranged so that their phase offset has a "$\pi/2$" difference, in another word, they are orthogonal to each other.

We write the output current $I_I(t)$ at the balanced receiver of the upper path as,

$$I_I(t) \propto \left|E_{sig}(t)\right|\left|E_{sig}(t-\tau)\right|\cos\left(\angle E_{sig}(t)-\angle E_{sig}(t-\tau)\right), \qquad (2.2.13)$$

where $\angle E_{sig}(t)-\angle E_{sig}(t-\tau)$ is the phase difference between two adjacent samples. Using this solely and having τ equal to the symbol period T_s one could demodulate a DBPSK signal as discussed in Section 2.2.2. However, for demodulating a DPSK / PSK signal with more than two phase states, a second path is required. With an additional "$\pi/2$" difference within the delay path, the output current can be written as

$$I_Q(t) \propto \left|E_{sig}(t)\right|\left|E_{sig}(t-\tau)\right|\sin\left(\angle E_{sig}(t) - \angle E_{sig}(t-\tau)\right). \tag{2.2.14}$$

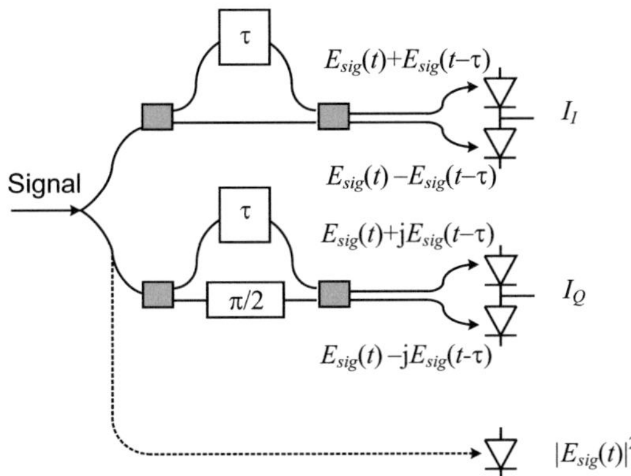

Fig. 2.18. Schematic of a self-coherent detector.

The inphase and quadrature phase signals are combined forming a complex signal,

$$u(t) = I_I(t) + jI_Q(t) = E_{sig}(t)E_{sig}^{\ *}(t-\tau). \tag{2.2.15}$$

In the DSP part of the receiver, the differential phase between adjacent samples can be calculated with $u(t)$,

$$\angle u(t) = \exp\left[j\left(\angle E_{sig}(t) - \angle E_{sig}(t-\tau)\right)\right] = u(t)/\left|u(t)\right|. \tag{2.2.16}$$

Assuming an arbitrary starting point $E_{sig}(t_0)$, the phase of a discrete signal $E_{sig}(t_0+n\tau)$ can be calculated as,

$$\angle E_{sig}(t_0 + n\tau) = \angle E_{sig}(t_0) + \sum_{m=1}^{n}\left(\angle u(t_0 + m\tau)\right) \tag{2.2.17}$$

The amplitude of $E_{sig}(t_0+n\tau)$ can be either detected with a separate photodiode (dashed branch in Fig. 2.18) [31], or by a geometric mean [32][33],

$$\left|E_{sig}(t_0 + n\tau)\right| \approx \left|u(t_0 + n\tau)\cdot u(t_0 + n\tau + \tau)\right|^{\frac{1}{2}} \tag{2.2.18}$$

As a result the electric field of the signal can be digitally recovered. A known phase error in the detector can be compensated by using electronic demodulator error compensation (EDEC) algorithm [33]. With a multi symbol phase estimation (MSPE) algorithm, a more accurate phase reference can be achieved in the phase recovery process [33][54]. In this way the detection penalty of differential phase shift keying (DPSK) compared to coherent detection of PSK signals can be substantially reduced. With the field recovery algorithm, electronic chromatic dispersion compensation can be performed via a multi stage FIR (finite impulse response) filter [33].

The geometric mean method for amplitude recovery can only be used for detection of DPSK/PSK signals with constant modulus, e.g., DBPSK, DQPSK, and D8PSK. The separate amplitude branch method can be used to detect multi-level signals including 8QAM and 16QAM [34]. It needs to be mentioned that a modification in the transmitter is necessary for phase pre-integration. A method to avoid using separate amplitude branch for detection of multi-level signals will be presented in Section 4.3.

The polarization diverse receiver can be easily achieved by positioning a polarization beam splitter (PBS) at the inputs of two self-coherent detection receivers, as shown in Fig. 2.19.

Because the input signal experiences random change of polarization states during transmission in the optical fiber link, a receiver technique for polarization demultiplexing is required. Similar as the differential direct detection, the polarization demultiplexing can be either done in the optical domain with an optical polarization controller or with other complicated optical and electronic means. In this thesis we do not require any additional optical hardware but use advanced DSP algorithm to perform the polarization demultiplexing. This will be further discussed in Section 4.2 and 4.3.

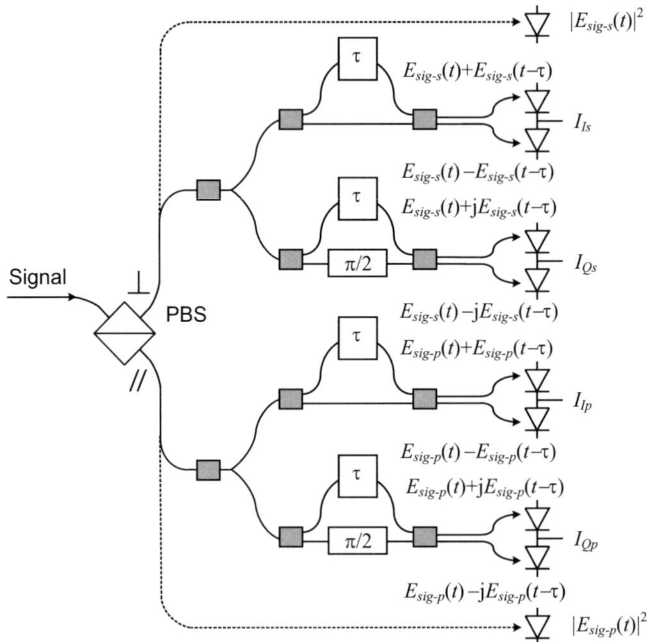

Fig. **2.19**. Schematic of a polarization diverse self-coherent detector.

2.3 Polarization Rotation on Optical Interface

Advanced receiver techniques are all based on interference. There are several conditions for optical beams to form interference [55]:

- Spatial and temporal overlap of the two fields
- Coherence of the two fields
- Non-orthogonal polarization states

Due to the fact that the optical frontend of the receiver discussed in this thesis is based on free-space optics technology, the first condition is relied on the proper design of the optical paths in the optical frontend. The second condition is met by using differential direct / self-coherent detection where the signal interferes with its delayed copy and the optical path difference is much smaller than the coherence length of the laser. In case of coherent detection the local oscillator has to be sufficiently close to the frequency of the signal. However, the third condition depends on various optical components in the optical frontend. In this section we will focus on how the optical interfaces of the components affect the polarization state of the optical signal.

2.3.1 Model of Single Optical Interface

The electric vector of a light beam towards an interface can be mapped on two orthogonal axes $\vec{E}_s$ and $\vec{E}_p$, Fig. 2.20. The incident beam ($\vec{E}_{i_s}$, $\vec{E}_{i_p}$) at interface between two different mediums results in a reflected beam ($\vec{E}_{r_s}$, $\vec{E}_{r_p}$) and a refracted beam ($\vec{E}_{t_s}$, $\vec{E}_{t_p}$). $\vec{E}_{i,r,t_s}$ is perpendicular to the plane formed by the incident direction and the normal to the interface. $\vec{E}_{i,r,t_p}$ is in parallel with the plane. In the following, $\vec{E}_s$ is called s light and $\vec{E}_p$ is p light.

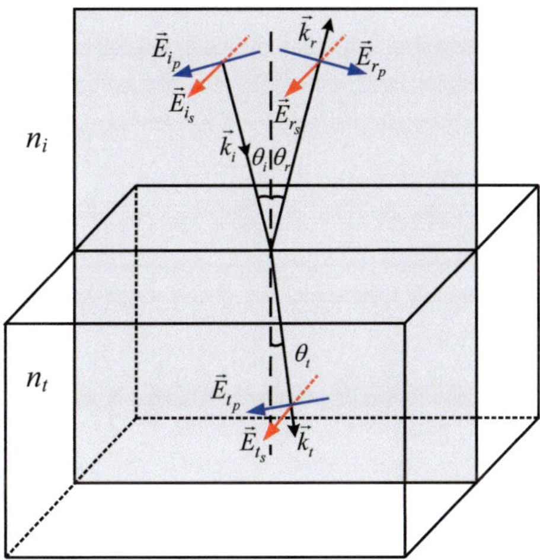

Fig. 2.20. Light incidents on an optical interface from a medium with refractive index of n_i to a medium with refractive index of n_t.

Given the refractive indices of mediums around interface n_i and n_t, applying Fresnel equations [55], the reflection coefficient $r_{s,p}$ and transmission coefficient $t_{s,p}$ for s and p lights at the interface are calculated to be

$$r_s = \frac{E_{r_s}}{E_{i_s}} = \frac{n_i \cos(\theta_i) - n_t \cos(\theta_t)}{n_i \cos(\theta_i) + n_t \cos(\theta_t)}, \quad r_p = \frac{E_{r_p}}{E_{i_p}} = \frac{n_t \cos(\theta_i) - n_i \cos(\theta_t)}{n_t \cos(\theta_i) + n_i \cos(\theta_t)}$$

$$t_s = \frac{E_{t_s}}{E_{i_s}} = \frac{2n_i \cos(\theta_i)}{n_i \cos(\theta_i) + n_t \cos(\theta_t)}, \quad t_p = \frac{E_{t_p}}{E_{i_p}} = \frac{2n_i \cos(\theta_i)}{n_t \cos(\theta_i) + n_i \cos(\theta_t)}.$$

$$(2.3.1)$$

where θ_i and θ_t are the incident and refractive angles (the angle between the direction vector of the incident / transmitted beam and the surface normal) around the optical interface. $E_{i,r,t_{s,p}}$ denote the electric fields of the incident, reflected and refracted beams.

We know from Snell's law that for an optical beam between two different mediums with different refractive indices the refractive angle is not equal to the incident angle except when the incident angle is $0°$ or $90°$. As a result the beam-waist normal to the direction vector (in parallel with Poynting vector) of the light is different for incident and refracted (transmitted) beams. Accordingly, the power, which is the average energy per unit time crossing the unit area normal to the Poynting vector of the incident, reflected and transmitted beams, can be calculated as $I_i\, A_{eff}\cos\theta_i$, $I_r\, A_{eff}\cos\theta_r$ and $I_t\, A_{eff}\cos\theta_t$, where $I_{i,r,t}$ is intensity of the beams and A_{eff} is the area of the interface and $\theta_{i,r,t}$ is the incident, reflective and refractive angle [55].

Intensity can be calculated as in Eq. (2.2.5). For convenience we write it again here,

$$I_{i,r,t} = \left\langle E_{i,r,t} \times H_{i,r,t} \right\rangle = \sqrt{\frac{\varepsilon_{i,r,t}\varepsilon_0}{\mu_{i,r,t}\mu_0}} \frac{1}{T} \int_0^T \left| E_{i,r,t}(t) \right|^2 dt = \frac{1}{2}\sqrt{\frac{\varepsilon_{i,r,t}\varepsilon_0}{\mu_{i,r,t}\mu_0}} \left| E_{i,r,t}(t) \right|^2, \qquad (2.3.2)$$

where ε_0 and μ_0 are electric permittivity and magnetic permeability of the vacuum, $\varepsilon_{i,r,t}$ and $\mu_{i,r,t}$ are relative electric permittivity and relative magnetic permeability of the mediums. Because the incident and reflective angle are identical and the incident and reflected beams are in the same medium, we have $\cos\theta_i = \cos\theta_r$, $\mu_i = \mu_r$, and $\varepsilon_i = \varepsilon_r$. The reflectance $R_{s,p}$ which represents the power ratio between the reflected and incident beams can be calculated as,

$$R_{s,p} = \left| \frac{I_{r_{s,p}} A_{eff} \cos\theta_r}{I_{i_{s,p}} A_{eff} \cos\theta_i} \right| = \left| \frac{I_{r_{s,p}}}{I_{i_{s,p}}} \right| = \left| \frac{\sqrt{\mu_i \varepsilon_r}}{\sqrt{\mu_r \varepsilon_i}} \frac{E_{r_{s,p}}}{E_{i_{s,p}}} \right|^2 = \left| r_{s,p} \right|^2. \qquad (2.3.3)$$

The transmittance $T_{s,p}$ which represents the power ratio between the transmitted and incident beams can be calculated as,

$$T_{s,p} = \left| \frac{I_{t_{s,p}} A_{eff} \cos\theta_t}{I_{i_{s,p}} A_{eff} \cos\theta_i} \right| = \left| \left(\frac{\sqrt{\mu_i \varepsilon_t}}{\sqrt{\mu_t \varepsilon_i}} \frac{\cos\theta_t}{\cos\theta_i} \right) \left| t_{s,p} \right|^2 \right|. \qquad (2.3.4)$$

For non-magnetic mediums, the relative permeability $\mu_{i,r,t} \approx 1$, and the refractive index of the mediums $n_{i,r,t} \approx \sqrt{\varepsilon_{i,r,t}}$. The above equation can be written as,

$$T_{s,p} = \left| \left(\frac{\sqrt{n_t} \cos\theta_t}{\sqrt{n_i} \cos\theta_i} \right) \left| t_{s,p} \right|^2 \right|. \qquad (2.3.5)$$

The angle of the complex coefficients $r_{s,p}$ and $t_{s,p}$ represents phase shift that the reflected and transmitted beams experience at the interface. The phase retardation is defined as the phase shift difference between the p and s lights,

$$\Delta\varphi_t = \angle t_p - \angle t_s, \quad \Delta\varphi_r = \angle r_p - \angle r_s, \tag{2.3.6}$$

where $\angle x$ denotes for the phase of a complex quantity x.

The phase retardation and the power ratio between the p and s light indicate the degree of polarization rotation for the reflected and transmitted beams compared to the incident one.

Now we use these equations to analyze the polarization rotation at different optical interfaces. Typical optical interfaces in an optical interferometer include dielectric to dielectric, and dielectric to metallic. For a dielectric to dielectric interface, we can further classify it into a less dense medium to dense medium interface and a dense medium to less dense medium interface. Two simple examples are air to glass and glass to air interfaces. For the dielectric to metallic interface, an example is glass to silver. The refractive indices of the mediums are listed in the following table.

Medium	Refractive index	Relative permeability
Glass	1.52	1
Air	1	1
Silver	0.14-11.3j	1

Table 2.2. Materials used in simulation.

For the definition of the clarity (left or right circular polarized state) we use the viewpoint of an observer that looks towards the field.

First, we take an air to glass interface for example. The power reflectance, transmittance and phase retardation of the beams versus the incident angle are plotted in Fig. 2.21.

In the plot, it can be seen that the power transmittance and reflectance vary with the change of incident angle. Especially, the power ratio between the s and p light of both reflected and transmitted beam changes with the increasing of the incident angle. The phase retardation of the transmitted light stays at '0°' at any incident angle, while the phase retardation of the reflected beam equal to '±180°' till the Brewster angle, then changes to '0°'. Therefore if the input light is a linearly polarized light that only consists of s or p components, the reflected and transmitted beams would stay at the same polarization state as the incident beam. However, if the input consists of non-zero p and s components, the polarization of the reflected and transmitted beams would change with the change of the incident angle.

For example, we have a 45° linearly polarized beam (s and p light have same power) coming from air to glass, by increasing the incident angle from 0° to 90°, the polarization of the reflected beam stays as linear polarization, however, rotates from −45° (orthogonal with the input light) to 45° (same as the input light), as shown in Fig. 2.22. This can be understood with the help of Fig. 2.21. For the reflected light, due to the 'π' phase difference between the p and s lights, at 0° incidence, the polarization of the reflected beam is orthogonal with the 45° polarized input beam. When increasing the incident angle, the power ratio between the s and p

light is increasing. At Brewster angle, p light disappears and afterwards the power ratio between the s and p lights start decreasing till 90° incident angle, the beam in fact does not hit the interface therefore it stays at the same polarization state as the incident one.

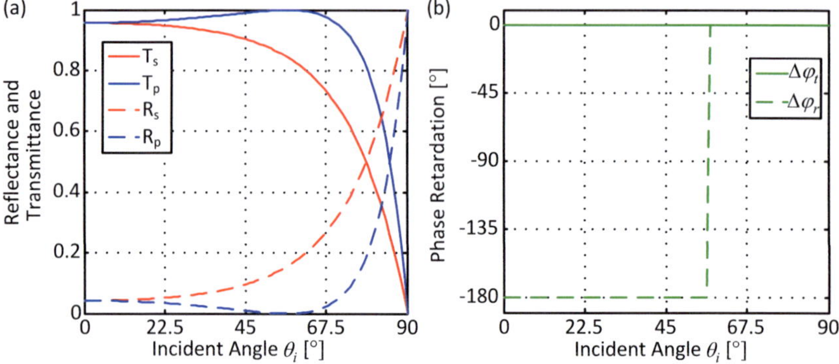

Fig. 2.21. (a) Power reflectance and transmittance of s and p light versus incident angle θ_i at optical interface from air to glass; (b) Phase retardation of the transmitted and reflected beams at optical interface from air to glass. Phase retardation is defined as the phase shift of p light minus the phase shift of s light.

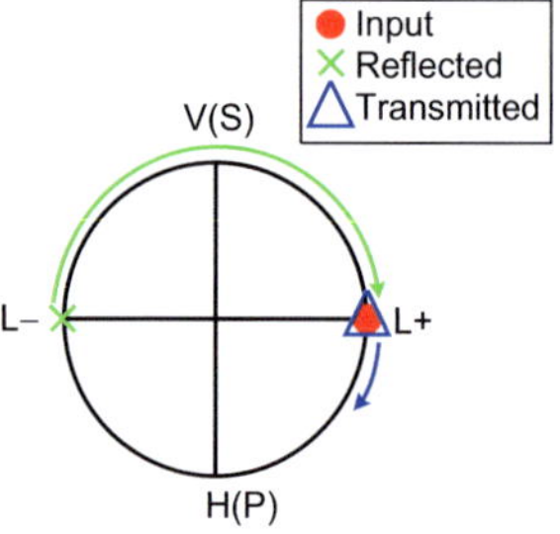

Fig. 2.22. Polarization rotation depends on the incident angle when a 45° linearly polarized input beam incident at an air-glass interface. L+: 45° linear polarization state; L−: −45° linear polarization state; V: 90° linear polarization state (s light); H: 0° linear polarization state (p light).

The polarization state of the transmitted light would also stay linearly polarized, however, rotate slightly from the polarization state of the incident beam with increasing incident angle. It can be also understood with the help of Fig. 2.21, the power ratio between p and s beams varies with increasing incident angle till there is no transmission at 90° incidence. And the phase retardation between p and s lights stays zero.

Now we take another example that the light comes from glass to air ($n_i > n_t$), the power reflectance, transmittance and phase retardation of the beams versus incident angle are plotted as in Fig. 2.23. This case is more complicated than the air-glass interface, because for both the reflected and transmitted beams the phase retardation between p and s light are no more 'π' or '0' once the incident angle goes beyond the critical angle. In case of a linearly polarized light having non-zero s and p components incident at the optical interface, such phase retardation between p and s light would result in an elliptical polarization state for the reflected and the transmitted beams.

We now analyze the glass to silver interface. Optical beam comes from glass to silver. The silver layer is thick enough to reflect most of the light back to glass. The rest is absorbed in the silver layer. The power reflectance, transmittance and phase retardation versus incident angle are plotted as in Fig. 2.24.

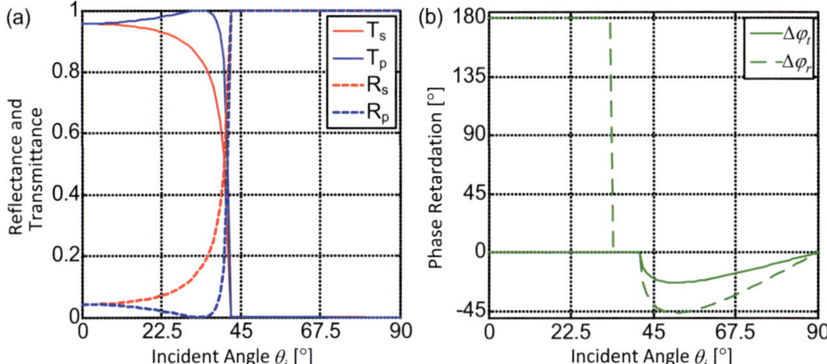

Fig. 2.23. (a) Power reflectance and transmittance of s and p light versus incident angle θ_i at optical interface from glass to air; (b) Phase retardation of the transmitted and reflected beams at optical interface from glass to air. Phase retardation is defined as the phase shift of p light minus the phase shift of s light.

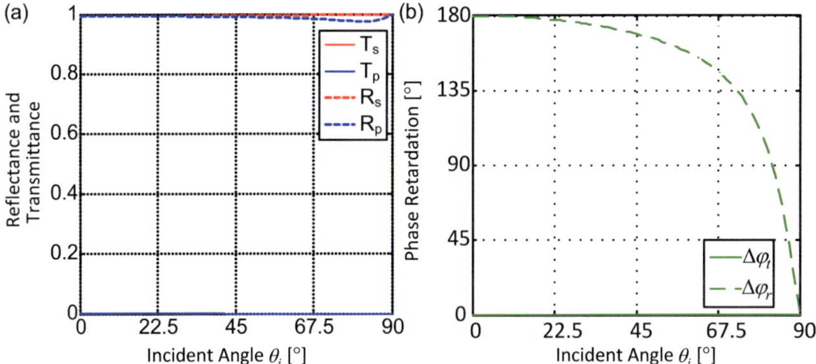

Fig. 2.24. (a) Power reflectance and transmittance of s and p light versus incident angle θ_i at optical interface from glass to silver; (b) Phase retardation of the transmitted and reflected beams at optical interface from glass to silver. Phase retardation is defined as the phase shift of p light minus the phase shift of s light.

It can be noticed that the power transmittance of the transmitted beam stays at '0' and the power reflectance of reflected beam is close to '1'. The power ratio between s and p light of the reflected beam stays almost constant. The phase retardation of the reflected beam is changing with increasing incident angle. Therefore similarly as the case of dense dielectric medium to less dense dielectric medium interface, a polarized beam with non-zero s and p components incident at the interface would result in a reflected beam with elliptical polarization state. The phase retardation of the transmitted beam stays at '0'. Because transmitted light is absorbed in the metallic layer, its polarization state is out of discussion.

As a conclusion, in all three cases, for both the reflected and transmitted beams, depending on the polarization state of the input beam, refractive indices of the mediums, and incident angle, the polarization states of the reflected and transmitted beam could change compared to the incident beam.

In addition there exist multi-layer thin films interface. In the next section, we will present a model of multi-layer optical interface.

2.3.2 Model of Multi–layer Optical Interface

Following derivation in [55], we now extend the model described in Eq. (2.3.1)–(2.3.4) to optical interface consisting of multi-layer thin films. We define that each layer consists of two boundaries, namely 'I' and 'II'. We firstly discuss for s light then for p light.

s Light

We plot the electromagnetic fields around one thin film layer as in Fig. 2.25.

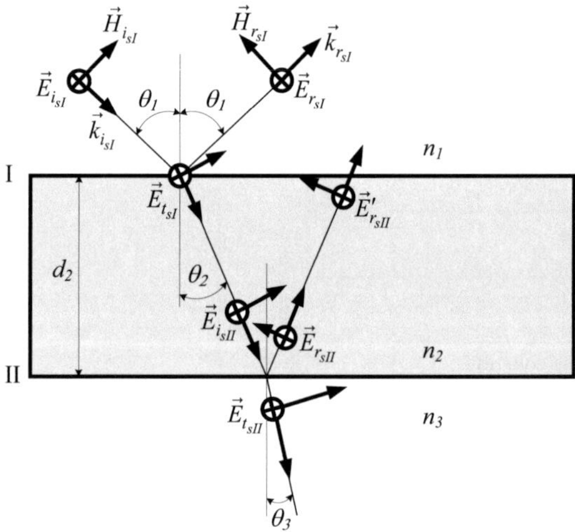

Fig. 2.25. Electromagnetic fields of s light at two boundaries of one thin film layer.

At boundary 'I' we have incident beam denoted with its electric vector $\vec{E}_{i_{sI}}$, magnetic vector $\vec{H}_{i_{sI}}$, and propagation direction $\vec{k}_{i_{sI}}$, reflected beam ($\vec{E}_{r_{sI}}$, $\vec{H}_{r_{sI}}$, $\vec{k}_{r_{sI}}$) in the first medium (refractive index is n_1), and transmitted beam ($\vec{E}_{t_{sI}}$, $\vec{H}_{t_{sI}}$, $\vec{k}_{t_{sI}}$) and back reflected beam from boundary 'II' ($\vec{E}'_{r_{sII}}$, $\vec{H}'_{r_{sII}}$, $\vec{k}'_{r_{sII}}$) in the second medium (n_2). Assuming all the mediums are isotropic, non-birefringent and nonmagnetic. The boundary condition requires that the tangential components of both the electric (E) and magnetic ($H=B/\mu$) fields be continuous across the boundaries. So at boundary 'I' we have,

$$E_{sI} = E_{i_{sI}} + E_{r_{sI}} = E_{t_{sI}} + E'_{r_{sII}},$$

$$H_{sI} = \cos\theta_1 H_{i_{sI}} - \cos\theta_1 H_{r_{sI}} = \cos\theta_2 H_{t_{sI}} - \cos\theta_2 H'_{r_{sII}},$$

(2.3.7)

where $E_{i,r,t_{s,p,I,II}}$ and $H_{i,r,t_{s,p,I,II}}$ denote the electric and magnetic fields of the incident, reflected and refracted beams of s and p lights at boundary I and II. Given $H = \sqrt{\varepsilon/\mu}E$ we further write H_{sI} as,

$$H_{sI} = \sqrt{\frac{\varepsilon_1\varepsilon_0}{\mu_1\mu_0}}\cos\theta_1\left(E_{i_{sI}} - E_{r_{sI}}\right) = \sqrt{\frac{\varepsilon_2\varepsilon_0}{\mu_2\mu_0}}\cos\theta_2\left(E_{t_{sI}} - E'_{r_{sII}}\right),$$

(2.3.8)

where ε_0 and μ_0 are electric permittivity and magnetic permeability of vacuum, $\varepsilon_{1,2}$ and $\mu_{1,2}$ are relative electric permittivity and relative magnetic permeability of the mediums.

At boundary 'II' we have incident beam ($E_{i_{sII}}$, $H_{i_{sII}}$, $k_{i_{sII}}$), reflected beam ($E_{r_{sII}}$, $H_{r_{sII}}$, $k_{r_{sII}}$) in the second medium (n_2), and transmitted beam ($E_{t_{sII}}$, $H_{t_{sII}}$, $k_{t_{sII}}$) in the third medium (n_3). According to the boundary condition, we have,

$$E_{sII} = E_{i_{sII}} + E_{r_{sII}} = E_{t_{sII}},$$

$$H_{sII} = \sqrt{\frac{\varepsilon_2\varepsilon_0}{\mu_2\mu_0}}\cos\theta_2\left(E_{i_{sII}} - E_{r_{sII}}\right) = \sqrt{\frac{\varepsilon_3\varepsilon_0}{\mu_3\mu_0}}\cos\theta_3 E_{t_{sII}}.$$

(2.3.9)

Within the film, we have,

$$E_{i_{sII}} = E_{t_{sI}}e^{-jk_0h_2}, \quad E_{r_{sII}} = E'_{r_{sII}}e^{jk_0h_2}.$$

(2.3.10)

Here k_0 is the propagation constant, and h_2 is the optical distance that the wave front travels between the two boundaries, e.g., in this case $h_2=n_2 d_2 \cos\theta_2$, given d_2 is the thickness of the optical thin film between the two boundaries. To get a relation between the electromagnetic fields at the two boundaries, we now put Eq. (2.3.10) into Eq. (2.3.9),

$$E_{sII} = E_{t_{sI}}e^{-jk_0h_2} + E'_{r_{sII}}e^{jk_0h_2} = E_{t_{sII}},$$

$$H_{sII} = \sqrt{\frac{\varepsilon_2\varepsilon_0}{\mu_2\mu_0}}\cos\theta_2\left(E_{t_{sI}}e^{-jk_0h_2} - E'_{r_{sII}}e^{jk_0h_2}\right) = \sqrt{\frac{\varepsilon_3\varepsilon_0}{\mu_3\mu_0}}\cos\theta_3 E_{t_{sII}}.$$

(2.3.11)

By reorganizing the equations, we have,

$$E_{t_{sI}} = \frac{1}{2}\left(E_{sII} + H_{sII}/\Upsilon_{s2}\right)e^{jk_0h_2},$$

$$E'_{r_{sII}} = \frac{1}{2}\left(E_{sII} - H_{sII}/\Upsilon_{s2}\right)e^{-jk_0h_2}, \text{ where } \Upsilon_{s2} = \sqrt{\frac{\varepsilon_2\varepsilon_0}{\mu_2\mu_0}}\cos\theta_2.$$

(2.3.12)

We further put Eq. (2.3.12) into Eq. (2.3.7),

$$E_{sI} = E_{t_{sI}} + E'_{r_{sII}} = \cos(k_0h_2)E_{sII} + j\frac{\sin(k_0h_2)}{\Upsilon_{s2}}H_{sII},$$

$$H_{sI} = \Upsilon_{s2}\left(E_{t_{sI}} - E'_{r_{sII}}\right) = j\Upsilon_{s2}\sin(k_0h_2)E_{sII} + \cos(k_0h_2)H_{sII}.$$

(2.3.13)

The relation between $[E_{sI}; H_{sI}]$ and $[E_{sII}; H_{sII}]$ can be written as,

$$\begin{bmatrix} E_{sI} \\ H_{sI} \end{bmatrix} = \mathcal{M}_2 \begin{bmatrix} E_{sII} \\ H_{sII} \end{bmatrix}, \text{ where } \mathcal{M}_2 = \begin{bmatrix} \cos(k_0 h_2) & j\dfrac{\sin(k_0 h_2)}{\Upsilon_{s2}} \\ j\Upsilon_{s2}\sin(k_0 h_2) & \cos(k_0 h_2) \end{bmatrix}. \tag{2.3.14}$$

Therefore the electromagnetic fields at the two boundaries are related by a characteristic matrix $\mathcal{M}_2$. Here the index number '2' in $\mathcal{M}_2$ is same as the index of the layer. Now, we assume that we have multiple layers, that for each layer the characteristic matrix is

$$\mathcal{M}_i = \begin{bmatrix} \cos(k_0 h_i) & j\dfrac{\sin(k_0 h_i)}{\Upsilon_{si}} \\ j\Upsilon_{si}\sin(k_0 h_i) & \cos(k_0 h_i) \end{bmatrix}, \text{ where } \Upsilon_{si} = \sqrt{\dfrac{\varepsilon_i \varepsilon_0}{\mu_i \mu_0}}\cos\theta_i. \tag{2.3.15}$$

We have h_i as the optical path length within the layer 'i', ε_i as relative electric permittivity of the layer 'i', μ_i as the relative magnetic permeability of the layer 'i', and θ_i as the refractive angle from layer 'i−1' to the layer 'i'.

For a multi-layer thin film coating, the overall characteristic matrix is the product of the matrices of all the layers. Thus, we can write the relation between the electromagnetic field $[E_{sI}; H_{sI}]$ at the incident boundary of the multi-layer coating and the electromagnetic field $[E_{s(L+1)}; H_{s(L+1)}]$ at the final transmitted boundary of the multi-layer coating as,

$$\begin{bmatrix} E_{sI} \\ H_{sI} \end{bmatrix} = \mathcal{M} \begin{bmatrix} E_{s(L+1)} \\ H_{s(L+1)} \end{bmatrix}, \text{ with } \mathcal{M} = \mathcal{M}_2 \mathcal{M}_3 ... \mathcal{M}_L = \begin{bmatrix} m_{11} & m_{12} \\ m_{21} & m_{22} \end{bmatrix}. \tag{2.3.16}$$

Now to calculate the overall reflectivity and transmissivity coefficient of the thin films, we put Eq. (2.3.7) into this equation,

$$\begin{bmatrix} E_{sI} \\ H_{sI} \end{bmatrix} = \begin{bmatrix} E_{i_{sI}} + E_{r_{sI}} \\ \Upsilon_{s1}\left(E_{i_{sI}} - E_{r_{sI}}\right) \end{bmatrix} = \mathcal{M} \begin{bmatrix} E_{t_{s(L+1)}} \\ \Upsilon_{s(L+1)}E_{t_{s(L+1)}} \end{bmatrix}. \tag{2.3.17}$$

Therefore the overall transmissivity and reflectivity coefficient of the thin films for s light can be derived as,

$$\begin{aligned} t_s &= \frac{2\Upsilon_{s1}}{m_{11}\Upsilon_{s1} + m_{12}\Upsilon_{s1}\Upsilon_{s(L+1)} + m_{21} + m_{22}\Upsilon_{s(L+1)}}, \\[2mm] r_s &= \frac{m_{11}\Upsilon_{s1} + m_{12}\Upsilon_{s1}\Upsilon_{s(L+1)} - m_{21} - m_{22}\Upsilon_{s(L+1)}}{m_{11}\Upsilon_{s1} + m_{12}\Upsilon_{s1}\Upsilon_{s(L+1)} + m_{21} + m_{22}\Upsilon_{s(L+1)}}. \end{aligned} \tag{2.3.18}$$

Assuming all the mediums are nonmagnetic, the refractive index of the layer 'i' $n_i = \sqrt{\varepsilon_i \mu_i} \approx \sqrt{\varepsilon_i}$. The power transmittance and reflectance can be calculated similarly as in Eq. (2.3.3) and (2.3.5),

$$T_s = \left|\left(\frac{n_{L+1}\cos\theta_{L+1}}{n_I \cos\theta_1}\right)|t_s|^2\right|, \quad R_s = |r_s|^2. \tag{2.3.19}$$

When $L=1$, this is same as the equations for single interface.

p Light

Similarly as for *s* light, we depict the electromagnetic fields around the boundaries of one layer of the thin film in Fig. 2.26.

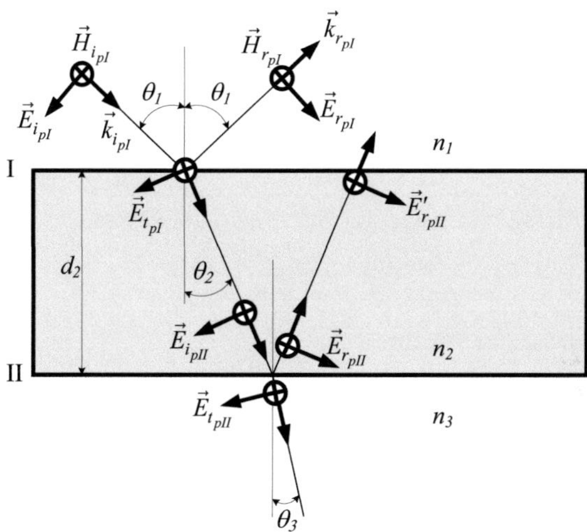

Fig. 2.26. Electromagnetic fields of *p* light at two boundaries of one thin film layer.

At boundary 'I' we have,

$$E_{pI} = \cos\theta_1 E_{i_{pI}} - \cos\theta_1 E_{r_{pI}} = \cos\theta_2 E_{t_{pI}} - \cos\theta_2 E'_{r_{pII}},$$

$$H_{pI} = \sqrt{\frac{\varepsilon_1 \varepsilon_0}{\mu_1 \mu_0}} \left(E_{i_{pI}} + E_{r_{pI}} \right) = \sqrt{\frac{\varepsilon_2 \varepsilon_0}{\mu_2 \mu_0}} \left(E_{t_{pI}} + E'_{r_{pII}} \right). \tag{2.3.20}$$

At boundary 'II' we have,

$$E_{pII} = \cos\theta_2 E_{i_{pII}} - \cos\theta_2 E_{r_{pII}} = \cos\theta_3 E_{t_{pII}},$$

$$H_{pII} = \sqrt{\frac{\varepsilon_2 \varepsilon_0}{\mu_2 \mu_0}} \left(E_{i_{pII}} + E_{r_{pII}} \right) = \sqrt{\frac{\varepsilon_3 \varepsilon_0}{\mu_3 \mu_0}} E_{t_{pII}}. \tag{2.3.21}$$

In the thin film layer we have,

$$E_{i_{pII}} = E_{t_{pI}} e^{-jk_0 h_2}, \quad E_{r_{pII}} = E'_{r_{pII}} e^{jk_0 h_2}. \tag{2.3.22}$$

Similarly as for *s* light, we have the relation between the two electromagnetic fields as,

$$\begin{bmatrix} E_{pI} \\ H_{pI} \end{bmatrix} = \mathcal{M}_2 \begin{bmatrix} E_{pII} \\ H_{pII} \end{bmatrix}, \text{ with } \mathcal{M}_2 = \begin{bmatrix} \cos(k_0 h_2) & j\dfrac{\sin(k_0 h_2)}{\Upsilon_{p2}} \\ j\Upsilon_{p2}\sin(k_0 h_2) & \cos(k_0 h_2) \end{bmatrix}, \Upsilon_{p2} = \frac{1}{\cos\theta_2}\sqrt{\frac{\varepsilon_2 \varepsilon_0}{\mu_2 \mu_0}}. \tag{2.3.23}$$

For multi-layer thin film coating, we have $\mathcal{M}_i$,

$$\mathcal{M}_i = \begin{bmatrix} \cos(k_0 h_i) & j\dfrac{\sin(k_0 h_i)}{\Upsilon_{pi}} \\ j\Upsilon_{pi}\sin(k_0 h_i) & \cos(k_0 h_i) \end{bmatrix}, \text{ where } \Upsilon_{pi} = \frac{1}{\cos\theta_i}\sqrt{\frac{\varepsilon_i\varepsilon_0}{\mu_i\mu_0}}. \tag{2.3.24}$$

The relation between the two electromagnetic fields is,

$$\begin{bmatrix} E_{pl} \\ H_{pl} \end{bmatrix} = \mathcal{M}\begin{bmatrix} E_{p(L+1)} \\ H_{p(L+1)} \end{bmatrix}, \text{ with } \mathcal{M} = \mathcal{M}_2\mathcal{M}_3...\mathcal{M}_L = \begin{bmatrix} m_{11} & m_{12} \\ m_{21} & m_{22} \end{bmatrix}. \tag{2.3.25}$$

So the transmissivity and reflectivity of p light can be calculated as,

$$t_p = \frac{2\Upsilon_{pl}}{m_{11}\Upsilon_{pl} + m_{12}\Upsilon_{pl}\Upsilon_{p(L+1)} + m_{21} + m_{22}\Upsilon_{p(L+1)}} \cdot \frac{\cos\theta_1}{\cos\theta_{L+1}},$$

$$r_p = \frac{-m_{11}\Upsilon_{pl} - m_{12}\Upsilon_{pl}\Upsilon_{p(L+1)} + m_{21} + m_{22}\Upsilon_{p(L+1)}}{m_{11}\Upsilon_{pl} + m_{12}\Upsilon_{pl}\Upsilon_{p(L+1)} + m_{21} + m_{22}\Upsilon_{p(L+1)}}. \tag{2.3.26}$$

The power transmittance and reflectance are,

$$T_p = \left|\left(\frac{n_{L+1}\cos\theta_{L+1}}{n_1\cos\theta_1}\right)|t_p|^2\right|, \quad R_p = |r_p|^2. \tag{2.3.27}$$

Now we can use these equations to analyze some typical optical coatings used in the optical interferometer system.

Anti-reflection (AR) coating is widely used in free-space optics to enhance the transmission of the optical interfaces. It consists of multiple layers, and each layer has an optical distance $h_i = \lambda/4$. The most simple AR coating consists of only two layers, see Fig. 2.27, the refractive indices of the layers are designed under the following principle [55],

$$\left(\frac{n_H}{n_L}\right)^2 = \frac{n_s}{n_0}, \tag{2.3.28}$$

where n_H is the refractive index of the coating material with relatively higher refractive index, n_L the refractive index of the coating material with relatively lower refractive index, n_s the refractive index of the substrate, and n_0 the refractive index of the incidence medium.

In practice, it's difficult to find materials that exactly match Eq. (2.3.28). Materials with refractive indices close to the designed value are used to optimize the AR performance. In Table 2.3, we list a number of materials that are generally used in the AR coating.

Material	Refractive index
Air	1
Cerium Fluoride (CeF_3)	1.63
Zirconium Dioxide (ZrO_2)	2.08
Glass	1.52
Air	1
Magnesium Fluoride (MgF_2)	1.38

Material	Refractive index
Zinc Sulfide (ZnS)	1.48
Glass	1.52

Table 2.3. Refractive indices of materials used in two kinds of AR coating.

All these materials have relative permeability close to '1'.

We have glass as the substrate, air as the incidence medium, CeF_3 as L material and ZrO_2 as H material. Substituting the refractive indices into Eq.(2.3.28), we have the left side equal to 1.46 which is smaller than the right side that equals to 1.52.

The transmittance and phase retardation for p and s lights at the AR coating can be calculated with Eq. (2.3.19) and (2.3.27). The thickness of each layer is set to $\lambda/4$ at normal incidence. The transmittance of the beam traveling from air to the AR coating and then to glass is plotted in Fig. 2.28. The transmittance stays close to '1' at incidence angle $0°$–$10°$.

We then plot the phase retardation of the transmitted beam in Fig. 2.29. The phase retardation also stays close to '$0°$' at incidence angle $0°$–$10°$.

If the incident beam propagates in the reverse direction, we would have transmittance of a beam traveling from glass to the AR coating and then to air. A plot of the transmittance in this case is shown in Fig. 2.30. The phase retardation is also plotted in Fig. 2.31. The performance is similar as for the case from air to glass.

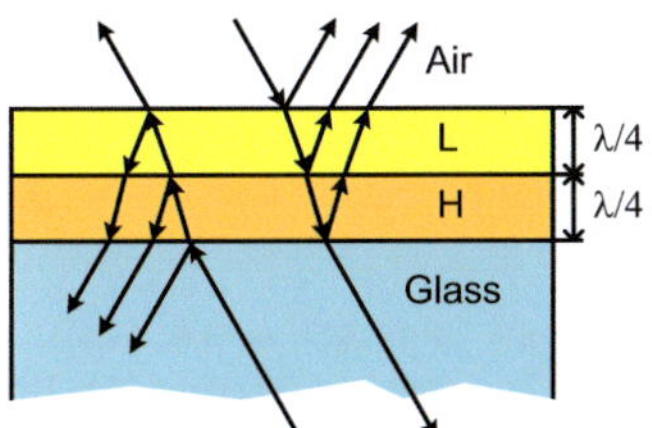

Fig. 2.27. AR coating consists of two layers, namely 'L' and 'H'. Incidence medium in this case is air and the substrate is glass.

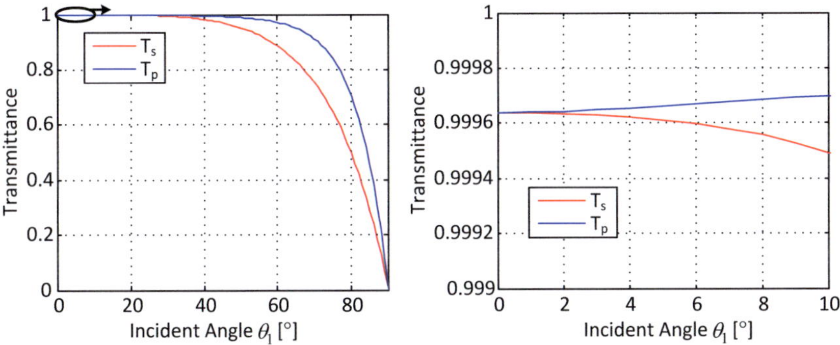

Fig. 2.28. The beam travels from air to glass with an AR coating consisting of CeF_3 and ZrO_2. Transmittance versus incident angle θ_1 from, left: $0°$ to $90°$; right: $0°$ to $10°$.

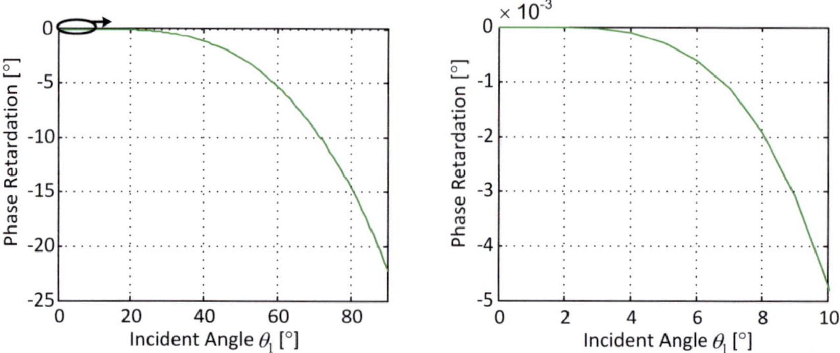

Fig. 2.29. The beam travels from air to glass with an AR coating consisting of CeF_3 and ZrO_2. Phase retardation of the transmitted beam versus incident angle θ_1 from, left: 0° to 90°; right: 0° to 10°.

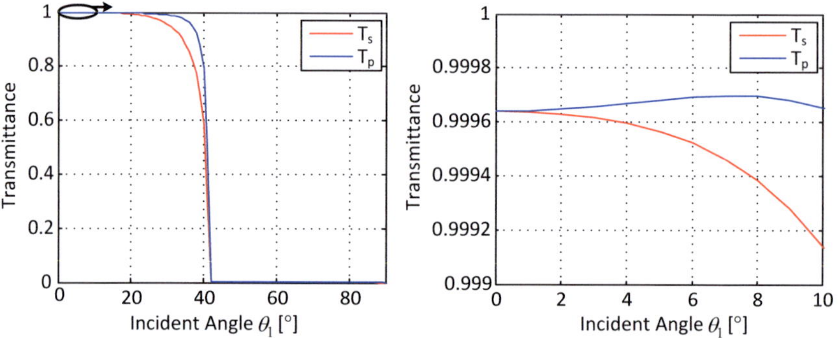

Fig. 2.30. The beam travels from glass to air with an AR coating consisting of CeF_3 and ZrO_2. Transmittance versus incident angle θ_1 from, left: 0° to 90°; right: 0° to 10°.

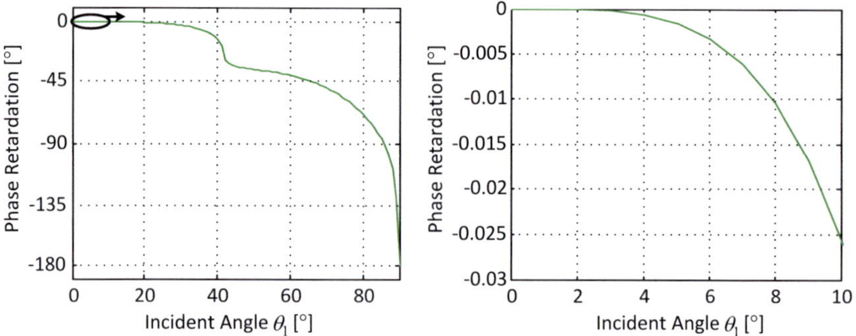

Fig. 2.31. The beam travels from glass to air with an AR coating consisting of CeF_3 and ZrO_2. Phase retardation of the transmitted beam versus incident angle θ_1 from, left: 0° to 90°; right: 0° to 10°.

In summary, when the incident angle is below 10°, the transmittance for both air to glass and glass to air is always above 99.8%, and the phase retardation is negligible. The AR coating causes negligible change to the polarization state of the transmitted beam compared to the incident beam.

Another example is using MgF_2 as L material and ZnS as H material. We have the left side of Eq. (2.3.28) equal to 1.15 which is much smaller than the right side that equals to 1.52. Again, the coating thickness is designed to a quarter wavelength at a 0° incidence angle.

Transmittance and phase retardation for a beam traveling from air to the AR coating and then to glass are plotted in Fig. 2.32 and in Fig. 2.33. They are also plotted for a beam travelling from glass to the AR coating and then to air in Fig. 2.34 and Fig. 2.35.

The transmittance is smaller than the first case due to the worse matched HL coating. However, the phase retardation is slightly better.

In summary, in the case of incident angle close to the designed value, the polarization of the transmitted light is not influenced by the AR coating.

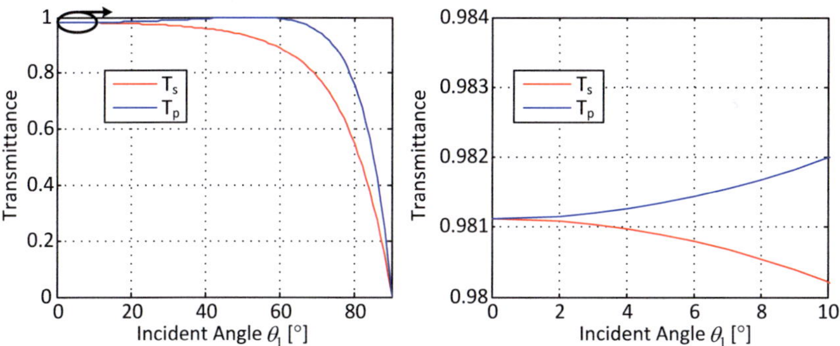

Fig. 2.32. The beam travels from air to glass with an AR coating consisting of ZnS and MgF_2. Transmittance versus incident angle θ_1 from, left: 0° to 90°; right: 0° to 10°.

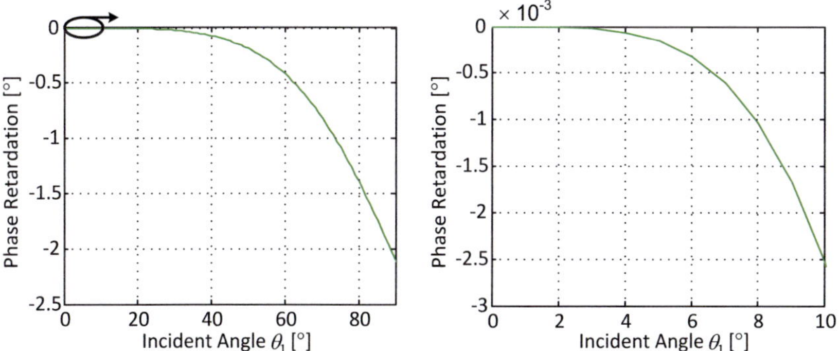

Fig. 2.33. The beam travels from air to glass with an AR coating consisting of ZnS and MgF_2. Phase retardation of the transmitted beam versus incident angle θ_1 from, left: 0° to 90°; right: 0° to 10°.

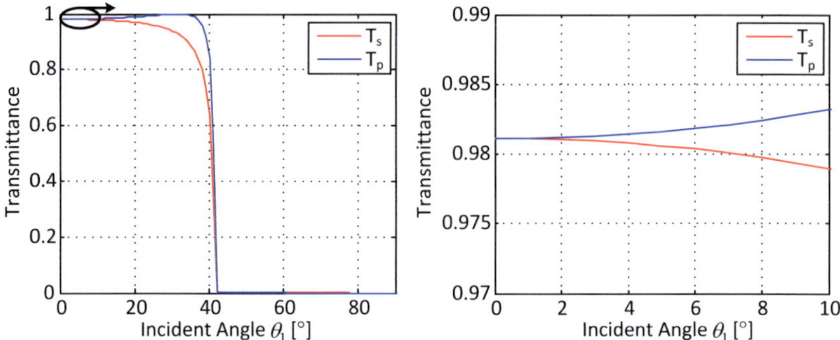

Fig. 2.34. The beam travels from glass to air with an AR coating consisting of ZnS and MgF$_2$. Transmittance versus incident angle θ_1 from, left: 0° to 90°; right: 0° to 10°.

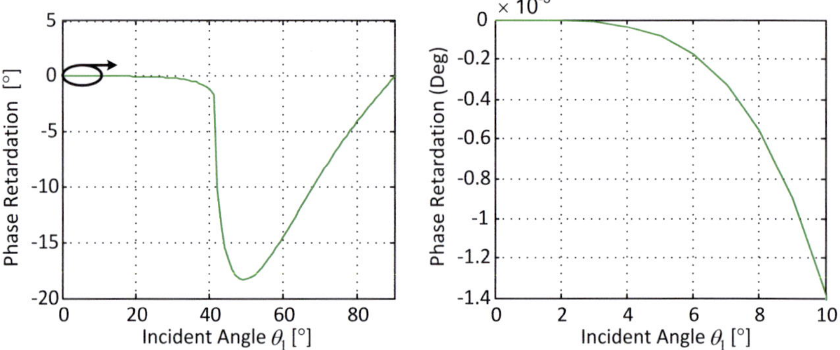

Fig. 2.35. The beam travels from glass to air with an AR coating consisting of ZnS and MgF$_2$. Phase retardation of the transmitted beam versus incident angle θ_1 from, left: 0° to 90°; right: 0° to 10°.

Non-polarizing beam splitting coating is used to split the input beam in two paths with equal power for both p and s lights. This generally requires multiple layers of thin film coatings. Here we present one example with 9 layers of thin film operating at wavelength of 650 nm. The coating structure is presented in Fig. 2.36. The materials used in the coating are listed in the Table 2.4.

Material	Refractive index	Thickness
Glass	1.52	-
MgF$_2$ (Magnesium Fluoride)	1.37	154.14 nm
TiO$_2$ (Titanium Dioxide)	2.31	94.09 nm
MgO (Magnesium Oxide)	1.72	190.06 nm
TiO$_2$ (Titanium Dioxide)	2.31	25.44 nm
MgF$_2$ (Magnesium Fluoride)	1.37	180.78 nm

Material	Refractive index	Thickness
TiO_2 (Titanium Dioxide)	2.31	64.03 nm
Ag (Silver)	0.14-11.4j	4.42 nm
TiO_2 (Titanium Dioxide)	2.31	59.68 nm
MgF_2 (Magnesium Fluoride)	1.37	197.95 nm
Glass	1.52	-

Table 2.4. Refractive indices and thickness of material used in non-polarizing beam splitting coating.

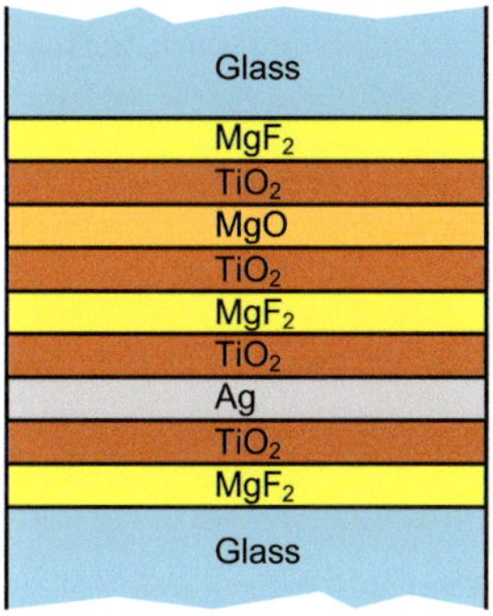

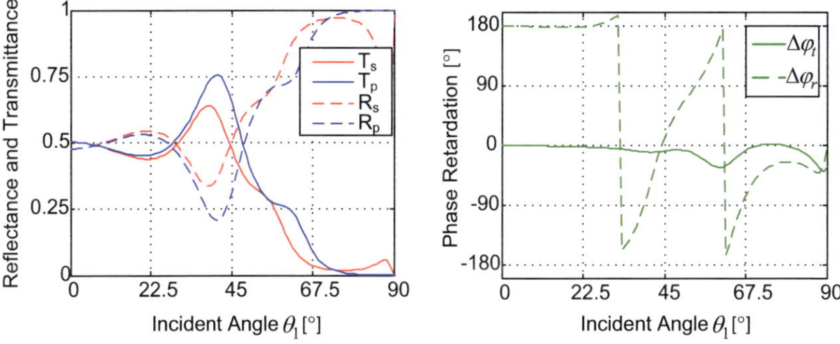

Fig. 2.36. Non-polarizing beam splitting coating consists of 9 layers of thin film.

The power transmittance and reflectance of *p* and *s* lights are plotted in the left side of Fig. 2.37, and the phase retardation of the transmitted and reflected beams are plotted in right side.

It can be noticed that around the incidence angle of 45°, the power splitting ratio for the transmitted and reflected beams are both close to 50% for both *s* and *p* lights. However, the phase retardations are all about ±10°. Therefore a polarization rotation would take place at the transmitted and reflected beams compared to the input beam if non-zero *s* and *p* components are carried.

Fig. 2.37. The beam travels from glass to glass with a non-polarizing beam splitting coating in between. Left: transmittance and reflectance versus incident angle θ_1 from 0° to 90°; right: phase retardations of the transmitted and reflected beams versus incident angle θ_1 from 0° to 90°.

2.4 Optical Elements used in Delay Interferometer

A free-space optical delay interferometer consists of collimators, reflectors, beam splitters and waveplates. In the following, each component will be briefly introduced.

2.4.1 Collimator

Optical collimators used in the interferometer system are to collimate the optical beams interfacing a fiber to free-space. The structure of a collimator is illustrated in Fig. 2.38. The beam coming out of the fiber first goes through a homogeneous medium which is called spacer. It then arrives at a piece of grin lens. The grin lens has a gradual varying refractive index. It functions as an objective lens which focuses the divergent beam and forwards it into free-space as a collimated beam. Anti-reflective coating can be applied on the front and back faces of the lens.

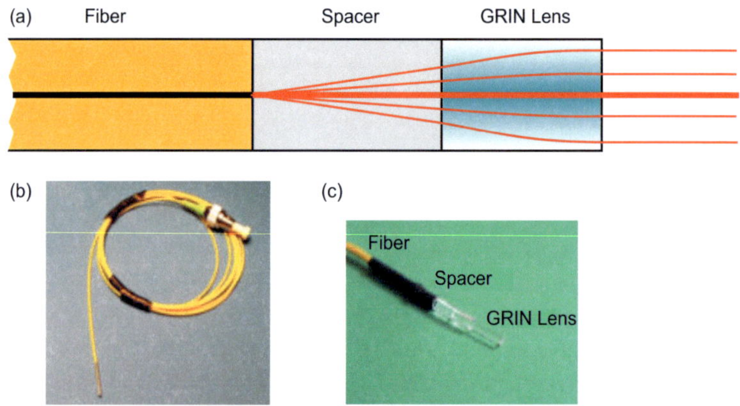

Fig. 2.38. (a) Schematic of a grin lens based optical collimator; (b) Photo of an optical collimator with a fiber pigtail (Grintech GmbH); (c) Photo of an optical collimator. (Grintech GmbH).

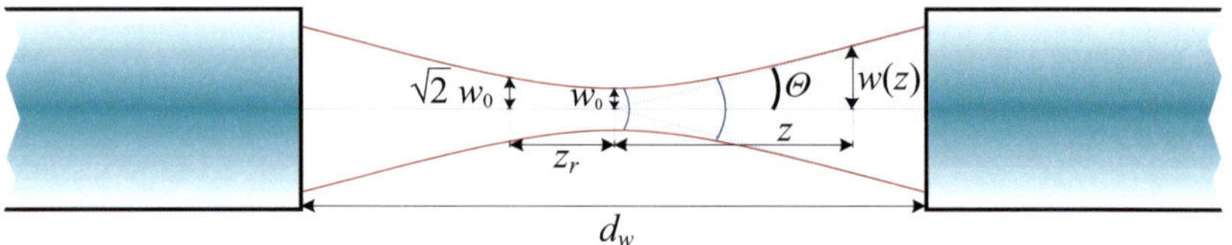

Fig. 2.39. Schematic of the Gaussian beam out of the optical collimators. w_0 is the beam waist, z_r the Rayleigh range, Θ the beam divergence angle, and d_w the working distance.

Normally, a fiber collimator lens is specified with a focal point for where the minimum beam waist would be. At the output of the collimator, the beam can be modeled as Gaussian beam that is defined with its beam waist w_0, beam divergence Θ and working distance d_w, see Fig. 2.39. The waist radius at position 'z' from the beam waist at '$z = 0$' is calculated as [56],

$$w^2(z) = w_0^2\left[1+\left(\frac{z}{z_r}\right)^2\right], \text{ where } z_r = \frac{\pi w_0^2}{\lambda}, \tag{2.4.1}$$

z_r is the Rayleigh range.

The working distance is two times of the distance between the collimator to the beam waist position, which is also the best reception position for a second collimator if said second collimator has the same characteristics as the input collimator.

2.4.2 Reflector

A reflector in an interferometer is used to fold the optical path and bring the beams to superimpose with each other. Various optical components can be utilized as a reflector. In this section we discuss the most popular ones, namely Porro prism and retro-reflector.

Porro Prism

A "Porro prism" - sometimes called a "right angle prism", has the shape of half a cubic which is cut along the diagonal face, see Fig. 2.40(a). In Fig. 2.40(b) a top view of the Porro prism is depicted. The incoming beam first hits the hypotenuse of the prism where an AR coating is usually attached. Then the beam hits on two orthogonal faces. The two optical surfaces can be either dielectric so that total internal reflection reflects the wave or metallic so that external refection reflects the wave. The Porro prism can be made of optical glass or comprise metallic mirrors with air in the center.

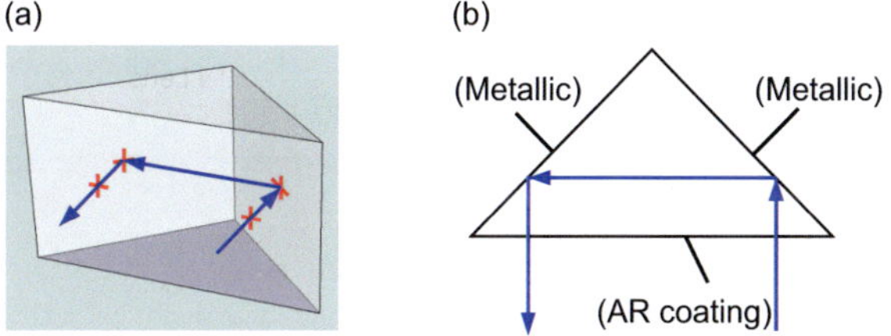

Fig. 2.40. (a) Schematic of a Porro prism; (b) Schematic of a Porro prism in top view.

We now discuss the polarization dependence of the Porro prism. First the incident beam passes the AR coating of the hypotenuse of the prism. As discussed in Section 2.3.2, an AR coating does not distort the polarization state of the signal when the incident angle is close to $0°$ as designed here. However, as shown in Fig. 2.23 and Fig. 2.24, both the total internal refection at a dielectric interface and the external refection at a metallic interface introduce significant phase retardation which could result in a polarization rotation for the reflected beam with respect to the incoming beam assuming non-zero s and p lights components. Compared to total internal reflection at a dielectric interface, the external reflection at a metallic interface causes less phase retardation and introduces negligible absorption. As a result, it's preferred to have metallic coating on the two orthogonal faces of the Porro prism.

To illustrate the difference between the dielectric and the metallic interface, we define an input beam comprising s and p lights,

$$\vec{E}_i = \begin{bmatrix} E_{i_s} \\ E_{i_p} \end{bmatrix}. \tag{2.4.2}$$

We define a transfer matrix for the optical interface number $i = 1$ or 2 of the two orthogonal faces by

$$\mathbf{M}_{r_i} = \begin{bmatrix} r_s & 0 \\ 0 & r_p \end{bmatrix}, \quad \mathbf{M}_{t_i} = \begin{bmatrix} t_s & 0 \\ 0 & t_p \end{bmatrix}, \tag{2.4.3}$$

where $\mathbf{M}_r$ is for the reflected beam and $\mathbf{M}_t$ the transmitted wave. The reflection coefficient ($r_{s,p}$) and transmission coefficient ($t_{s,p}$) for s and p lights at the interface are calculated using Eq. (2.3.1), (2.3.18) and (2.3.26). After several interfaces, neglecting the propagator, the electric fields at the output can be written as,

$$\vec{E}_{r,t} = \mathbf{M} \begin{bmatrix} E_{i_s} \\ E_{i_p} \end{bmatrix}, \text{ where } \mathbf{M} = \mathbf{M}_{r_n,t_n} \dots \mathbf{M}_{r_2,t_2} \mathbf{M}_{r_1,t_1}. \tag{2.4.4}$$

Take a Porro prism with dielectric interface for example, ignoring the hypotenuse which is AR coated, we have the refractive index of glass $n = 1.52$, the medium outside of the Porro prism as air and the incident angle $\theta_i = 45°$. The transfer matrix of two dielectric interfaces after reflection is,

$$\mathbf{M} = \mathbf{M}_{r_2}\mathbf{M}_{r_1} = \begin{bmatrix} e^{j(-80.5°)} & 0 \\ 0 & e^{j(-161°)} \end{bmatrix}. \tag{2.4.5}$$

The power of s and p lights is conserved and the phase retardation $\Delta\varphi$ between the p and s lights is about $-80.5°$. By carefully selecting the material of the prism, it can be used as a quarter waveplate (QWP) or even a half waveplate (HWP) [57].

The transfer matrix of two metallic interfaces (silver with the refractive index $n = 0.14-11.3j$) at 45° incidence is,

$$\mathbf{M} = \mathbf{M}_{r_2}\mathbf{M}_{r_1} = \begin{bmatrix} 0.9954e^{j(-21.6°)} & 0 \\ 0 & 0.9908e^{j(-43.3°)} \end{bmatrix}. \tag{2.4.6}$$

There is an absorption that is almost identical for the s and p lights, i.e. less than 1%. The phase retardation $\Delta\varphi$ between the p and s lights is only $-21.6°$ which is much smaller than in the case of dielectric interface.

Retro-reflector

A retro-reflector has three faces which are orthogonal to each other when looking at it from the input, see Fig. 2.41(a). This ensures that the incoming beam is always reflected back in parallel but opposite direction of the beam source. Compared to a Porro prism, a retro-reflector reduces the alignment tolerance by paying a penalty by an additional reflection. The prism can be made with either optical glass or metallic mirrors with air in the center. In the case that it is made with optical glass, the front face is usually AR coated, and the three orthogonal faces can be either dielectric interface or metallic interface, see Fig. 2.41(b).

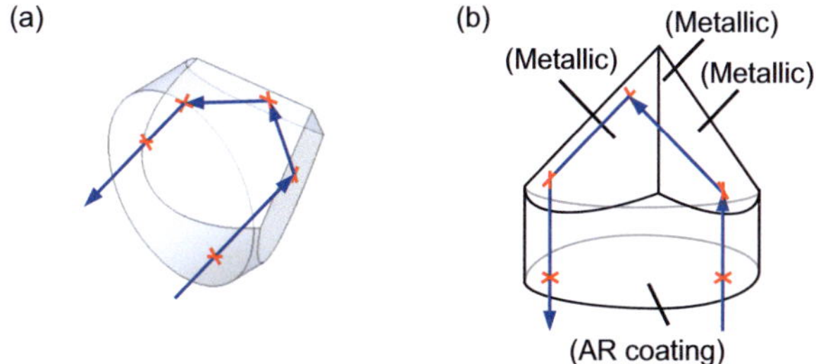

Fig. 2.41. (a) Schematic of a retro-reflector; (b) Schematic of the retro-reflector prism in side view.

To mathematically model a retro-reflector, we introduce a coordination system transfer matrix, because the beam no longer propagates in one plane. We first define the coordination system of the incident beam as three orthogonal vectors, which are the propagation vector $\vec{k}$, electric field vector of s and p lights $\vec{E}_{s,p}$. The propagation vector $\vec{k}$ is normal to the vector of electric flux density D and magnetic field B. Because the mediums are isotropic and we assume the beam to be TEM (transverse electromagnetic), $\vec{E}_{s,p}$ and $\vec{k}$ are orthogonal to each other. The interface is represented by its normal vector. As shown in Fig. 2.42, the angle between the propagation vector and normal vector is the incident angle θ_i.

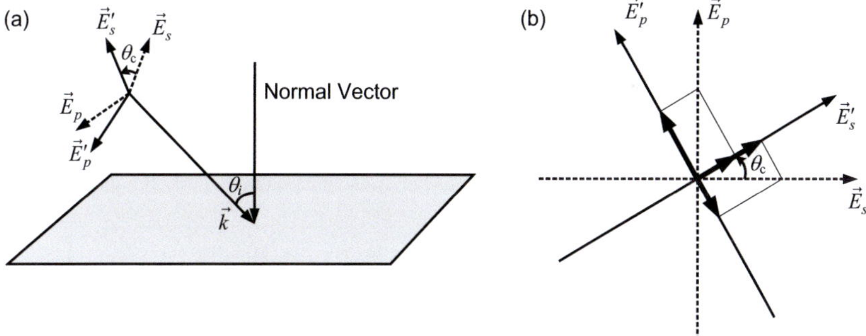

Fig. 2.42. Coordinate axes ($\vec{E}_{s,p}$ and $\vec{k}$) of the input has an angle of θ_c to its new coordination system ($\vec{E}'_{s,p}$ and $\vec{k}$) according to the current interface. (a) The propagation vector $\vec{k}$ has an incident angle θ_i with respect to the normal vector of the interface. (b) The electric vectors $\vec{E}_{s,p}$ are mapped onto $\vec{E}'_{s,p}$.

As shown in the figure, the incident beam previously has been represented with respect to the coordinate system along $\vec{E}_s$ and $\vec{E}_p$. This coordinate system is different from the new coordinate system with axes along $\vec{E}'_s$ and $\vec{E}'_p$. The two coordinate systems are offset with respect to each other by θ_c. A 2×2 transfer matrix **R** is used to map $\vec{E}_s$ and $\vec{E}_p$ onto $\vec{E}'_s$ and $\vec{E}'_p$. Here we define θ_c as the angle from $\vec{E}_s$ to $\vec{E}'_s$ counter-clockwise,

$$\begin{bmatrix} E'_s \\ E'_p \end{bmatrix} = \mathbf{R} \begin{bmatrix} E_s \\ E_p \end{bmatrix} = \begin{bmatrix} \cos(\theta_c) & \sin(\theta_c) \\ -\sin(\theta_c) & \cos(\theta_c) \end{bmatrix} \begin{bmatrix} E_s \\ E_p \end{bmatrix}. \qquad (2.4.7)$$

For retro-reflector with dielectric interfaces, assuming the incident beam normal incident at the front face, we define the initial $\vec{E}_s$ and $\vec{E}_p$ vectors according to the first glass to air interface that the beam encounters. We can calculate the incident angle θ_i to be 54.7°. By inserting the incident angle and refractive indices as listed in Table 2.2 into Eq. (2.3.1) and into the transfer matrix in Eq. (2.4.3). The transfer matrix then is,

$$\mathbf{M}_{r_1} = \begin{bmatrix} r_s & 0 \\ 0 & r_p \end{bmatrix} = \begin{bmatrix} e^{j(-79.8°)} & 0 \\ 0 & e^{j(-125.2°)} \end{bmatrix}. \qquad (2.4.8)$$

The offset angle θ_c between the coordination systems of the two glass to air interfaces is 60°. The transfer matrix from glass to air interface number 1 to 2 is calculated as in Eq. (2.4.7),

$$\mathbf{R}_{1\text{-}2} = \begin{bmatrix} \cos(\theta_c) & \sin(\theta_c) \\ -\sin(\theta_c) & \cos(\theta_c) \end{bmatrix} = \begin{bmatrix} 0.5 & 0.866 \\ -0.866 & 0.5 \end{bmatrix}. \qquad (2.4.9)$$

Similarly we can derive the transfer matrix of glass to air interface number 2 and 3, and the transfer matrix from interface number 2 to number 3 ($\theta_c = -60°$) and from number 3 back to the original states as at interface number 1 in an opposite direction ($\theta_c = 60°$). Neglecting the propagator, the final output can be calculated,

$$\begin{bmatrix} E_{r_s} \\ E_{r_p} \end{bmatrix} = \mathbf{R}_{3\text{-}1}\mathbf{M}_{r_3}\mathbf{R}_{2\text{-}3}\mathbf{M}_{r_2}\mathbf{R}_{1\text{-}2}\mathbf{M}_{r_1} \begin{bmatrix} E_{i_s} \\ E_{i_p} \end{bmatrix} = \begin{bmatrix} 0.68e^{j(-107°)} & 0.74e^{j(29.8°)} \\ 0.74e^{j(-104.8°)} & 0.68e^{j(-2°)} \end{bmatrix} \begin{bmatrix} E_{i_s} \\ E_{i_p} \end{bmatrix}. \qquad (2.4.10)$$

The s and p lights at the output are mixture product of the p and s lights at the input. Take a 45° linearly polarized input $\vec{E}_i = [1; 1]\exp(j\omega_0 t)$ for example, we have output equal to $[1.10e^{j(66.4°)}; 0.88e^{j(-56.5°)}]\exp(j\omega_0 t)$. The overall power is conserved, however, distributed unequally in s and p lights. The phase retardation is $-122.9°$ which results in an elliptical polarized output.

For the metallic coating, by taking refractive indices from Table 2.2 we can calculate the overall transfer matrix as

$$\mathbf{M} = \mathbf{R}_{3\text{-}1}\mathbf{M}_{r_3}\mathbf{R}_{2\text{-}3}\mathbf{M}_{r_2}\mathbf{R}_{1\text{-}2}\mathbf{M}_{r_1} = \begin{bmatrix} 0.99e^{j(-52.7°)} & 0.02e^{j(117.1°)} \\ 0.02e^{j(134.6°)} & 0.99e^{j(127.1°)} \end{bmatrix}. \qquad (2.4.11)$$

The cross-coupling between the s and p lights are negligible. The phase retardation introduced by the retro-reflector is about 'π' which results in flip of the polarization with respect to the input. Take a 45° linearly polarized input $\vec{E}_i = [1; 1]\exp(j\omega_0 t)$ for example, we now have an output of $[1.0083e^{j(-52.5°)}; 0.9693e^{j(127.3°)}]\exp(j\omega_0 t)$ which is orthogonal with the input. This is normal for a reflected beam after three reflections. An experimental verification of the above discussion can be found in [58].

As a conclusion, a retro-reflector with metallic coating is the best choice in terms of polarization conservation and tolerance of alignment. A Porro prism with metallic coating saving one reflection can be considered to be used in applications where dimensions and cost are critical.

2.4.3 Beam Splitter

Beam splitter is used in the interferometer system to split the input beam in two paths. It can be a polarization beam splitter that splits the input beam in two orthogonal polarizations or a non-polarizing beam splitter that splits the input equally for both polarizations. In Fig. 2.43, a schematic drawing of the beam splitter is depicted.

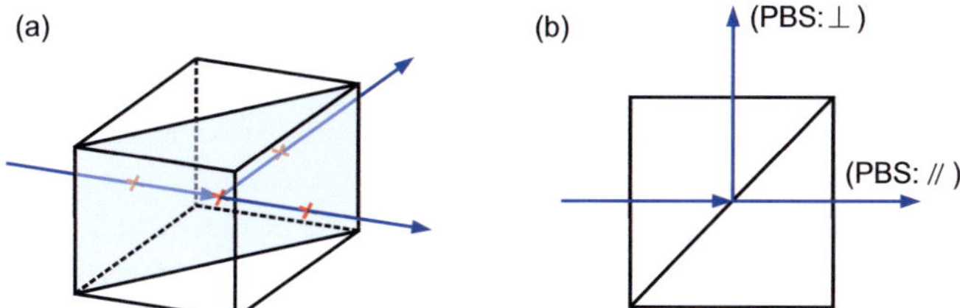

Fig. 2.43. (a) Schematic of a beam splitter; (b) Schematic of the beam splitter in top view.

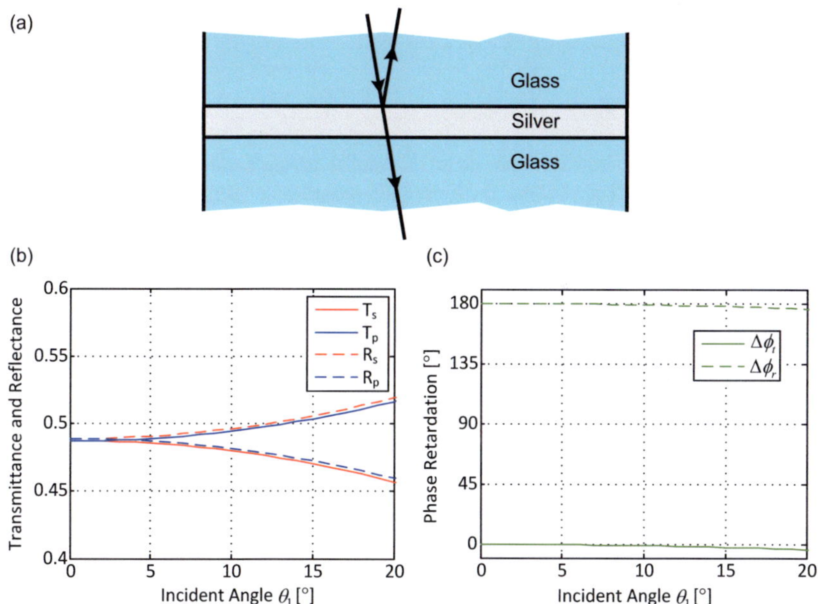

Fig. 2.44. (a) Beam incident from glass to a silver film of 5.7 nm thick and to another glass; (b) Power reflectance and transmittance of s and p lights versus incident angle θ_1; (c) Phase retardations between p and s light of the transmitted and reflected beams versus incident angle θ_1.

For a polarization beam splitter, the essential characteristic is the extinction ratio between the two polarizations at each output. For a non-polarizing beam splitter the splitting ratio for each polarization is of interest. Additionally, a possible polarization difference between the beams in the two paths is of interest for an interferometer.

Non-polarizing beam splitters can be fabricated with multi-layer thin films such as shown in Fig. 2.36 and Fig. 2.37. Though the power splitting ratio is almost 50:50 for both *s* and *p* lights, the phase retardations of the transmitted and reflected beams have a non-zero or non-π difference. Thus the polarization states of the reflected and the transmitted beams change differently with respect to the incident beam.

A second way to fabricate non-polarizing beam splitter is to use a very thin layer of metallic coating, e.g., a 5.7 nm thick thin silver film, in between of two glass mediums, see Fig. 2.44(a). The power transmittance and reflectance and phase retardation of the transmitted and reflected beam are presented in Fig. 2.44(b). When the incident angle is small, we could achieve a 50:50 splitting ratio for both *s* and *p* lights. The phase retardation of the transmitted beam is close to '0', and the phase retardation of the reflected beam is close to '180°'. Therefore it introduces negligible polarization change to the transmitted and reflected beams with respect to the incident one. However, such thin film is impractical for fabrication and the small incident angle is difficult for alignment.

2.4.4 Waveplate

Waveplates are also known as retarders which serve to introduce phase retardation between *p* and *s* lights thus change the polarization of an incident beam. A typical waveplate is a birefringent crystal whose orientation and thickness are carefully chosen. The crystal is cut so that the extraordinary axis or "optic axis" is parallel to the surfaces of the plate. Light polarized along this axis travels through the crystal at a different speed than light perpendicular with the axis, thus a phase difference is created. When the phase difference is equal to π, the waveplate is named as half-waveplate. When it's π/2, the waveplate is named as quarter waveplate [59].

A special waveplate is made with liquid crystal. Liquid crystal has the property that when there is external electric field, its molecules would rearrange their orientation which leads to a change of its birefringence, see Fig. 2.45. In this way a variable waveplate can be introduced.

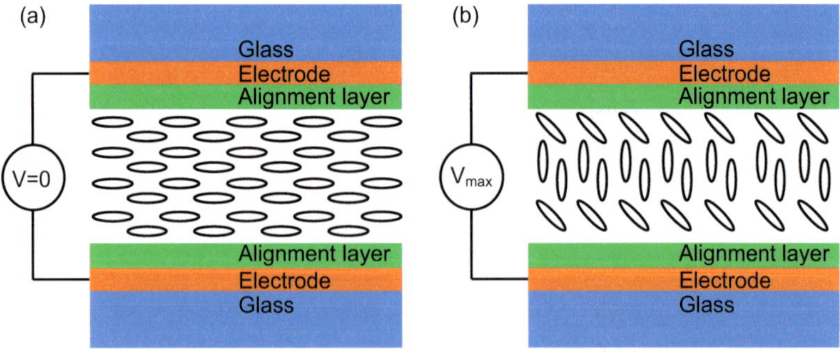

Fig. 2.45. (a) Liquid crystal molecules arrangement when there is no voltage applied; (b) Liquid crystal molecules arrangement when there is maximal voltage applied. (Picture is taken from Thorlabs website)

3 Configurations of Delay Interferometer

Based on the theoretical analysis of polarization rotation upon optical interfaces, delay interferometers (DI) including polarization insensitive DI and polarization and phase diverse DIs, comprising tunable time delay have been developed in this thesis.

In this chapter, we firstly present an optical delay interferometer characterized with tunable time delay and polarization insensitivity then introduce its control circuit. This part of the thesis has been published in Optics Express [J5]. In the second half of this chapter we present a design of a micro-optical interferometer that comprises four delay interferometers and two optical 90° hybrids. It can be operated either with self-coherent or coherent receiver principle. This part of the thesis has been filed as a patent [P1] and submitted to the Optics Express [J1].

3.1 Free-space Optical Delay Interferometer with Tunable Delay and Phase

The content of this section is a direct copy of the Journal publication [J5]:
J. Li, K. Worms, R. Maestle, D. Hillerkuss, W. Freude, and J. Leuthold, "Free-space optical delay interferometer with tunable delay and phase," Opt. Express, vol. 19, no. 12, pp. 11654-11666, 2011.
Minor changes have been done to adjust the color, line type and position of figures and notations of variables.

A free-space optical delay interferometer (DI) featuring a continuously tunable time delay, polarization insensitive operation with high extinction ratios and accurate phase and time delay monitoring scheme is reported. The polarization dependence is actively mitigated by adjusting a birefringent liquid-crystal device. The DI has been tested for reception of D(m)PSK signals.

3.1.1 Introduction

Delay interferometers (DIs) are needed to demodulate differential phase shift keying modulation formats, e.g., DBPSK, DQPSK and higher-level formats like D8PSK (for short, D(m)PSK). These formats have various advantages regarding noise, nonlinearities, and dispersion tolerance [4][6]. Recently, DIs have been used to detect the amplitude and optical phase of coherent data signals [31][33][C21]. The main advantage of such a "self-coherent" detection scheme [33] over a real coherent reception is that self-coherent detection does not require an expensive, possibly wideband-tunable local oscillator. Moreover, DIs have also been used as optical filters for all-optical wavelength conversion [60][J9]. More recently, DIs have been used as ultrafast optical FFT processing elements enabling FFT processing of a 10 Tbit/s OFDM signal [J7].

While delay interferometers have become important for all of the above and many more applications the fabrication of versatile delay interferometers is a challenge to this day. For the above mentioned applications, a versatile DI should feature:

- Tunability in delay time and tunability for phase. In almost any of the above mentioned applications a good reception quality, requires adaption of the time delay to the symbol rate [50]. In again other instances, it is sometimes advantageous to deviate from the one-symbol delay in order to mitigate transmission impairments caused by effects such bandwidth-narrowing by concatenated filters [61]. A delay interferometer with a continuously tunable delay would fulfill both requirements. Recently, a DI with adaptive delay has been presented by using cascaded Mach-Zehnder interferometers [62], however, this only provides a discrete switching of three delays. A continuously tunable DI has been proposed in [35], where the delay is introduced by passing two orthogonal polarizations through a tunable differential group delay (DGD) element. This scheme, however, requires that the input signal has equal power for both orthogonal polarizations.

- Low polarization-dependent loss (PDL), and especially a low polarization-dependent frequency shift (PDFS), which are of particular importance for demodulating D(m)PSK signals [63]. This is normally achieved by carefully selecting the optical coatings, which should perform for all polarizations alike.

- Accurate monitoring and control of the operating point defining phase and time delay. Phase monitoring with a fixed time delay has been intensively discussed based on RF power monitoring [64] or based on a correlation method [65]. Although the efficiency of both methods has been demonstrated, the techniques are complicated, and they are limited to certain data formats and bitrates. A pilot-tone driven lock-in algorithm can be much more effective in fixing the DI operating point, especially for DQPSK reception, when a DI pair needs be locked to the desired $\pi/2$ relative phase offset. Sophisticated time delay control techniques can be found in the field of distance metrology [66][67]. A simple solution would be to use a known pilot tone, and to count fringes when tuning the DI.

In this section we present a polarization-insensitive, continuously tunable free-space optical delay interferometer. Our approach operates over a broad wavelength range.

The section is organized as follows: We first model an optical delay interferometer mathematically. Then we present an appropriate free-space setup, which provides a calibrated, continuously tunable time delay from 0 ps to 100 ps. We achieve PDFS mitigation by employing a liquid-crystal device to compensate the birefringence within the system. Delay and phase are controlled with the help of a pilot tone. Experimentally, we demodulate DQPSK signals at 11.7 GBd, 28 GBd and 42.7 GBd, present eye diagrams and bit error ratios (BER), and compare the outcome to the results measured with a commercially available DI with fixed delay. In the last section we show that with two additional polarization beam splitters four logic DIs can be folded into one single DI which serves as a tunable polarization division multiplexing (PolMUX) receiver for polarization multiplexed D(m)PSK signals.

3.1.2 DInterferometer (DI) Modeling

A schematic DI layout is shown in Fig. 3.1. It consists of two couplers and two optical paths of different length. The upper path has a time delay of τ compared to the lower path. The input electric fields $E_{in,1}$ and $E_{in,2}$ are split by the first coupler ($\mathbf{S_I}$), then experience two different optical paths, and are then combined at the second coupler ($\mathbf{S_{II}}$), where two output fields $E_{out,1}$ and $E_{out,2}$ are generated.

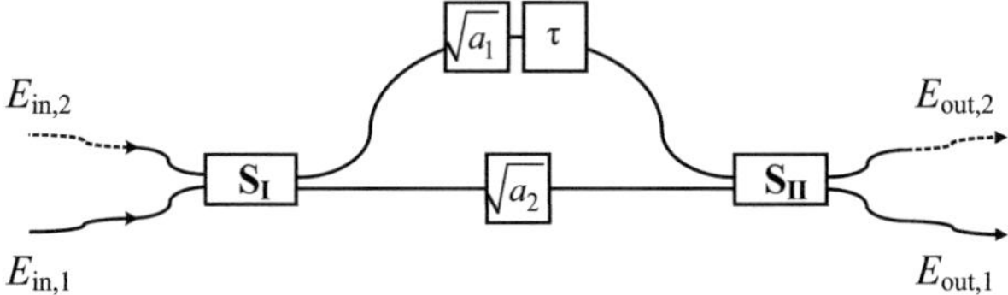

Fig. 3.1. Schematic of an optical delay interferometer (DI). Inputs ($E_{in,\,1}$ and $E_{in,\,2}$) are split by a coupler ($\mathbf{S_I}$) into two paths with a_1 and a_2 as the respective power loss factors. A time delay τ is introduced between the two paths. The signals interfere on the other coupler ($\mathbf{S_{II}}$) where two outputs ($E_{out,\,1}$ and $E_{out,\,2}$) are generated.

Following [60] we convolve (symbol $*$) the impulse response matrix of the DI with the time-dependent column matrix of input electric fields $E_{in,1,2}$, and find the time-dependent output electric fields $E_{out,1,2}$ (only one polarization is considered):

$$\begin{bmatrix} E_{out,2} \\ E_{out,1} \end{bmatrix} = (\mathbf{S_{II}} \cdot \mathbf{T} \cdot \mathbf{S_I}) * \begin{bmatrix} E_{in,2} \\ E_{in,1} \end{bmatrix}, \text{ with } \mathbf{S_I} = \begin{bmatrix} \sqrt{a_I(1-s_I)} & j\sqrt{a_I s_I}\exp(j\phi_{12}^I) \\ j\sqrt{a_I s_I}\exp(j\phi_{21}^I) & \sqrt{a_I(1-s_I)} \end{bmatrix},$$

$$\mathbf{T} = \begin{bmatrix} \sqrt{a_1}\delta(t-\tau) & 0 \\ 0 & \sqrt{a_2}\delta(t) \end{bmatrix}, \text{ and } \mathbf{S_{II}} = \begin{bmatrix} \sqrt{a_{II}(1-s_{II})} & j\sqrt{a_{II}s_{II}}\exp(j\phi_{12}^{II}) \\ j\sqrt{a_{II}s_{II}}\exp(j\phi_{21}^{II}) & \sqrt{a_{II}(1-s_{II})} \end{bmatrix}. \tag{3.1.1}$$

The matrices $\mathbf{S_I}$ and $\mathbf{S_{II}}$ in Eq. (3.1.1) describe the two couplers, where a phase factor j provides the ideal phase relations between the two output signals, which might be impaired by phase offsets ϕ_{12}^I, ϕ_{21}^I, ϕ_{12}^{II}, and ϕ_{21}^{II}. The quantities $a_I s_I$, $a_I(1-s_I)$, $a_{II}s_{II}$ and $a_{II}(1-s_{II})$ are the power splitting ratios of the two couplers in the form of amplitude factors. If $a_I = 1$ and $a_{II} = 1$, the couplers are lossless. The Dirac distributions in matrix $\mathbf{T}$ are the impulse responses of the "long" (upper) and the "short" (lower) arms having a group delay difference τ, see Fig. 3.1. Assuming a monochromatic optical signal with angular frequency $\omega_c = 2\pi f_c$, a change of τ introduces a phase offset $-\tau\omega_c$ in the upper arm. The quantities a_1, $a_2 < 1$ are the power loss factors in the two paths.

For a D(m)PSK signal demodulator with only one input to be used, i.e., $E_{in,1} \neq 0$, $E_{in,2} = 0$, Eq. (3.1.1) simplifies to

$$E_{out,1}(t) = \int_{-\infty}^{\infty} \left[-\sqrt{a_1 a_I a_{II} s_I s_{II}}\,\delta(x-\tau)\exp\left[j\left(\phi_{21}^{II} + \phi_{12}^I\right)\right] \right.$$
$$\left. + \sqrt{a_2 a_I a_{II}(1-s_I)(1-s_{II})}\,\delta(x)\right]E_{in,1}(t-x)\,dx, \tag{3.1.2}$$

$$E_{\text{out},2}(t) = j \int_{-\infty}^{\infty} \left[\sqrt{a_1 a_I a_{II} s_I (1-s_{II})} \, \delta(x-\tau) \exp\left(j\phi_{12}^I\right) \right.$$

$$\left. + \sqrt{a_2 a_I a_{II} s_{II} (1-s_I)} \, \delta(x) \exp\left(j\phi_{12}^{II}\right) \right] E_{\text{in},1}(t-x)\, dx.$$

We can also write Eq. (3.1.2) in the frequency domain. We define the Fourier transforms of $E_{\text{out},1,2}(t)$ and $E_{\text{in},1}(t)$ by $\mathcal{E}_{\text{out},1,2}(f)$ and $\mathcal{E}_{\text{in},1}(f)$, respectively. The associated transfer functions are $H_{1,2}(f) = |H_{1,2}(f)|\exp[j\Phi_{1,2}(f)] = \mathcal{E}_{\text{out},1,2}(f)/\mathcal{E}_{\text{in},1}(f)$.

The power transfer function then becomes

$$|H_1(f)|^2 = a_2 a_I a_{II} - \left[a_2 a_I a_{II} s_I + a_2 a_I a_{II} s_{II} - (a_1 + a_2) a_I a_{II} s_I s_{II} \right.$$

$$\left. + 2S \cos\left(2\pi f\tau - \phi_{21}^{II} - \phi_{12}^I\right) \right],$$

$$|H_2(f)|^2 = a_2 a_I a_{II} s_{II} + a_1 a_{II} s_I - (a_1 + a_2) a_I a_{II} s_I s_{II} \qquad (3.1.3)$$

$$+ 2S \cos\left(2\pi f\tau + \phi_{12}^{II} - \phi_{12}^I\right),$$

where $S = a_I a_{II} \sqrt{a_1 a_2 s_I s_{II} (1-s_I)(1-s_{II})}$.

For the associated phases we find:

$$\Phi_1(f) = \tan^{-1}\left[\frac{\sqrt{a_1 a_I a_{II} s_I s_{II}} \, \sin\left(2\pi f\tau - \phi_{21}^{II} - \phi_{12}^I\right)}{\sqrt{a_2 a_I a_{II}(1-s_I)(1-s_{II})} - \sqrt{a_1 a_I a_{II} s_I s_{II}} \, \cos\left(2\pi f\tau - \phi_{21}^{II} - \phi_{12}^I\right)} \right],$$

$$\Phi_2(f) = \tan^{-1}\left[\frac{\sqrt{a_1 a_I a_{II} s_I (1-s_{II})} \, \cos\left(2\pi f\tau - \phi_{12}^I\right) + \sqrt{a_2 a_I a_{II} s_{II} (1-s_I)} \, \cos\left(\phi_{12}^{II}\right)}{\sqrt{a_1 a_I a_{II} s_I (1-s_{II})} \, \sin\left(2\pi f\tau - \phi_{12}^I\right) - \sqrt{a_2 a_I a_{II} s_{II} (1-s_I)} \, \sin\left(\phi_{12}^{II}\right)} \right].$$

$$(3.1.4)$$

From the frequency dependence of the phase we derive the group delay $t_{g1,2}(f) = [-1/(2\pi)] \cdot [d\Phi_{1,2}(f)/d(f)]$

$$t_{g1}(f) = \tau \frac{a_1 a_I a_{II} s_I s_{II} - S\cos(2\pi f\tau - \phi_{21}^{II} - \phi_{12}^I)}{|H_1(f)|^2},$$

$$(3.1.5)$$

$$t_{g2}(f) = \tau \frac{a_1 a_I a_{II} s_I - \left[a_1 a_I a_{II} s_I s_{II} - S\cos(2\pi f\tau + \phi_{12}^{II} - \phi_{12}^I) \right]}{|H_2(f)|^2}.$$

For an illustration of Eq. (3.1.5) we consider two extreme cases: When $s_I, s_{II} = 0$, $a_I, a_{II} = 1$, the input field $E_{\text{in},1}$ in Fig. 3.1 passes exclusively through the lower arm to the output 1. The squared magnitude of the transfer function at the outputs is calculated to be $|H_1(f)|^2 = a_2$, $|H_2(f)|^2 = 0$, and the output group delay is $t_{g1}(f) = 0$ according to the assumption Eq. (3.1.1). When $s_I, s_{II} = 1$, $a_I, a_{II} = 1$, the input $E_{\text{in},1}$ in Fig. 3.1 propagates through the upper arm to output 1 without coupling into the lower arm. The squared magnitude of the transfer function at the outputs can be calculated, $|H_1(f)|^2 = a_1$, $|H_2(f)|^2 = 0$, and the associated output group delay is $t_{g1}(f) = \tau$ according to the assumption formulated in Eq. (3.1.1).

We describe an ideal DI (no loss, symmetric splitting) by the parameter set

$$a_1, a_2, a_I, a_{II} = 1, \quad s_I, s_{II} = \frac{1}{2}, \quad \text{and} \quad \phi_{12}^I, \phi_{21}^I, \phi_{12}^{II}, \phi_{21}^{II} = 0. \qquad (3.1.6)$$

We define f_0 as a frequency where destructive interference occurs at output 1, and constructive interference at output 2. We plot $|H_{1,2}(f)|^2$ as well as phases $\Phi_{1,2}(f)$ and group delays $t_{g1,2}(f)$ as a function of the frequency offset from f_0, blue curves in Fig. 3.2. The frequency offset is given in units of the free spectral range $FSR = 1/\tau$. The two outputs have complementary power transfer functions $|H_{1,2}(f)|^2 = 1/2 \cdot (1 \mp \cos(2\pi f \tau))$. This leads to an infinite extinction ratio $ER = (\text{maximum of output } 1, 2)/(\text{minimum of output } 2, 1)$, Fig. 3.2(a). Whenever destructive interference occurs and the transfer function $|H_{1,2}(f)|$ crosses zero, a π-phase shift is observed, Fig. 3.2(b). These phase jumps result in δ-functions of the group delay, $-\tau/2 \cdot \delta(f - N/2\tau)$, $N = 0, \pm1, \pm2, \pm3,\ldots$, while $t_{g1,2}(f) = \tau/2$ holds at all other frequencies, Fig. 3.2(c).

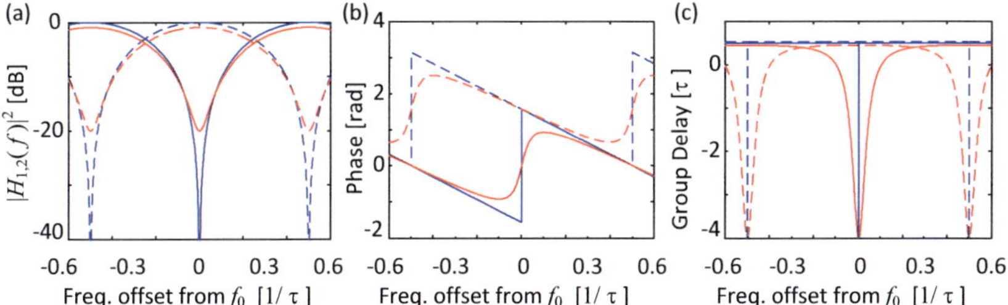

Fig. 3.2. Transfer function of the DI for different upper-arm power losses $a_1 = 1$ and $a_1 = 0.64$. The case $a_1 = 1$ denotes an ideal DI (blue). The case $a_1 = 0.64$ denotes an imbalanced DI with excess losses in the delayed arm (red). The three plots show (a) the power transfer functions $|H_{1,2}(f)|^2$, (b) the phase response $\Phi_{1,2}(f)$ and (c) the group delay $t_{g1,2}(f)$ for the constructive output port (dashed lines) and the destructive port (straight lines), respectively.

In practice, losses and splitting ratios in a DI are not ideal as defined in Eq. (3.1.6). As an example we assume 2 dB of losses in one arm, i.e., $a_1 = 0.64$, while all other parameters remain as in Eq. (3.1.6). For this case, the quantities $|H_{1,2}(f)|^2$, $\Phi_{1,2}(f)$ and $t_{g1,2}(f)$ are plotted in Fig. 3.2 (red curves). In contrast to the ideal case, the red curves have a finite $ER = 20$ dB, and an additional loss of 0.085 dB. Furthermore, instead of the π-phase jumps, the phase slopes are reduced near the frequency for destructive interference, which results in a finite negative group delay, Fig. 3.2(c). This behavior is typical for any deviation of a_1, a_2, a_I, a_{II}, and s_I, s_{II} from the ideal conditions defined in Eq. (3.1.6). Imperfect phase offsets ϕ_{12}^I, ϕ_{21}^I, ϕ_{12}^{II}, ϕ_{21}^{II} will introduce a frequency shift of the transfer functions.

In the following we use the general model of Eq. (3.1.1) for designing a practical implementation of a free-space optics delay interferometer.

3.1.3 DI Implementation and Polarization Dependence

DIs typically can be implemented as asymmetric Mach-Zehnder interferometers or as Michelson Interferometers. The two configurations are different though.

A Mach-Zehnder delay interferometer configuration is displayed in Fig. 3.3(a) [C30]. The interferometer is formed by a non-polarizing beam splitter (NPBS) and a reflector. The NPBS splits the collimated input signal in two paths. For changing the optical path length difference, the reflector is mounted on a micro-actuator to create a variable path length with an associated delay τ. However, we notice that the upper, longer path usually has more loss than the lower path due to the additional reflector, so that the extinction ratio is degraded. Moreover, the minimum time delay difference τ between the two paths is also limited by the fact that the upper optical path is always longer.

Conversely, the Michelson interferometer allows a zero optical path difference and minimal power difference between the signals in both arms, thus providing a (theoretically) infinite large free spectral range (FSR) in combination with a high ER at "Output 1, 2", see Fig. 3.3(b) [C19].

A prototype has been built as depicted in Fig. 3.3(c). The signals are coupled with fiber grin lenses into and out of the DI. The dimensions of $85 \times 45 \times 25$ mm^3 can be further decreased by using the LIGA technology as in [C30]. The actuator used in this setup has a mechanical step size of about 2 nm, which corresponds to a phase offset of about 1°. The mechanical tuning range amounts to 15 mm corresponding to a time delay tuning range of 100 ps. As the actuator responds within milliseconds, the tuning can be performed extremely fast.

Now we consider an input with two orthogonal polarization states, s and p ("s" perpendicular light with respect to the incidence plane, and "p" parallel light with respect to the incident plane). Because each surface in the system has polarization-dependent transmissivity and reflectivity, the transfer function of the DI is seriously affected by the input signal polarization state. While polarization dependent reflectivity can be minimized by carefully selected coatings, there remains a polarization-dependent phase shift. This phase offset results in a polarization dependent frequency shift (PDFS) of the DI transfer function.

To measure the PDFS of a DI, one could align the input signal parallel to one of two orthogonal (linear) polarizations, and calculate the difference between the respective transfer functions. In practice, because a DI is usually coupled to a standard single mode fiber, the input polarization state for the DI cannot be known. Therefore, we insert a polarizer directly after the input lens and record the spectral response of the DI for p polarization. Then by rotating the polarizer we record the spectral response for an s polarization. The spectrally shifted responses are found at "Output 1, 2" and are depicted in Fig. 3.4(a). Then with the knowledge of the PDFS, we adjust a birefringent element (by voltage tuning a liquid crystal (LC), Fig. 3.4(b)) for compensating the polarization-induced phase offset. When applying 1 V to the LC, we observe the spectral response in Fig. 3.4(b), where PDFS has been obviously mitigated as compared to Fig. 3.4(a).

The PDFS measurement in Fig. 3.4(c) shows how the birefringence can be undone by detuning the liquid crystal. We have plotted $\Delta\phi_{\text{pol}} = 2\pi\,(PDFS\,/\,FSR)$ against the voltage applied to the liquid crystal. The resolution of the plot is limited by the resolution of the equipment that allowed us to sweep the wavelength with steps of 10 pm only. For a FSR of 200 GHz (1.6 nm) we can resolve a phase difference of 2.2° (0.61% of FSR).

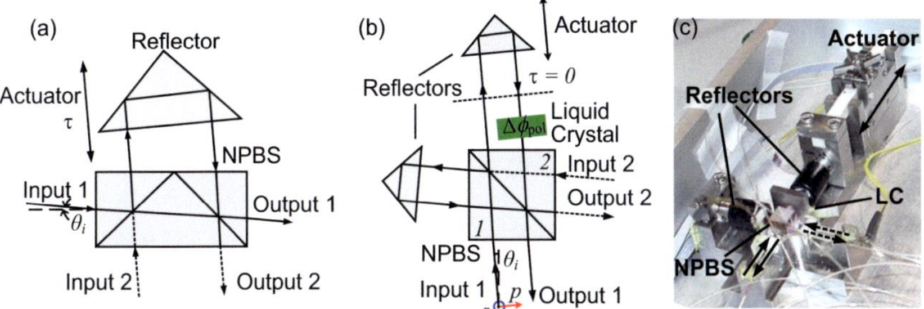

Fig. 3.3. Practical implementation of DI. (a) Schematic of the Mach-Zehnder-DI with two non-polarizing beam splitters (NPBS) combined in one unit, and one reflector. (b) Schematic of the Michelson DI with single NPBS and two reflectors. A liquid crystal (LC) compensates a polarization dependent frequency shift. (c) Photograph of a prototype of the Michelson DI.

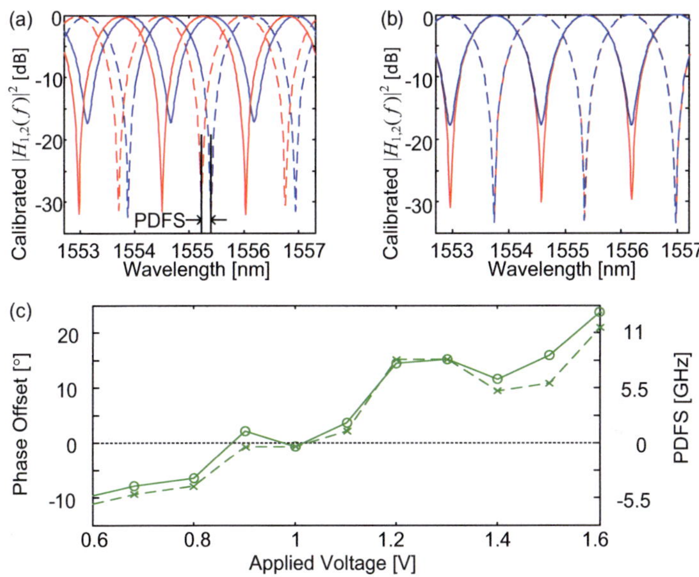

Fig. 3.4. Spectral response for *s* (blue) and *p* (red) polarizations. (a) Large PDFS, (b) LC is used to undo birefringence, (c) PDFS of DI for different voltages applied onto the LC (the precision of the measured offset phase is limited by the resolution of the measurement equipment). The plots show the output 2 (dashed lines) and output 1 (straight lines).

Once the birefringence is undone, it is undone for any delay τ as an additional free space does not add to the birefringence. In Fig. 3.4(a), (b) one can notice that the $ER = $ (maximum of output 1, 2) / (minimum of output 1, 2) at "Output 2" for both polarizations is larger than 30 dB. At "Output 1" and for *p*-polarization the ER is 30 dB as well, however, for the *s*-polarization only 18 dB is measured. This difference is due to the fact transmissivities and reflectivities of the NPBS coating are slightly polarization dependent. In our case output 2 has an almost ideal extinction ratio because the two beams constructively interfering in the beam splitter and being mapped onto output 2 undergo one reflection and one transmission each. However, the two beams interfering in the NPBS and being mapped to output 1 undergo two reflections and two transmissions, respectively, which leads to a power imbalance (splitting ratio difference) for the slightest imbalance in the transmission and reflection coefficients of the NPBS coating. As we discussed above, this will lead to the ER difference between the two outputs at two orthogonal polarizations. As commercially available fixed-delay DIs have typically *ER* > 18 dB, this prototype has comparable performance. A *PDFS* < (0.61% of *FSR*) is also good enough for DQPSK demodulation [50][63].

Polarization dependent loss (PDL) and differential group delay (DGD) are important as well. Commercial measurement instruments, e.g., Agilent 86038B Photonic Dispersion and Loss Analyzer, Agilent N7788BD Benchtop Optical Component Analyzer, usually provide such data. It is straight forward since PDL and DGD can be derived from the ratio of the transfer functions at orthogonal polarizations,

$$PDL = \left| 10\log_{10}\left(\left. |H_s(f)|^2 \middle/ |H_p(f)|^2 \right. \right) \right|,$$

$$DGD = \left| t_{g-s}(f) - t_{g-p}(f) \right|.$$

$$(3.1.7)$$

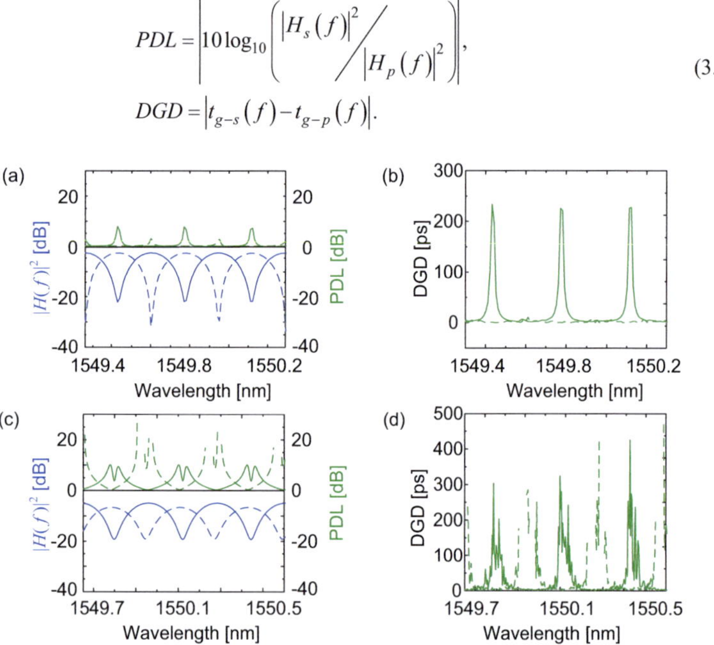

Fig. 3.5. Measured average squared magnitude of the transfer function for all possible polarizations (blue), of PDL (green), and of DGD (green) at DI output 1 (solid line) and 2 (dashed line); *FSR* ≈ 42.7 GHz. For minimized PDFS: (a) average loss and PDL, (b) DGD versus wavelength. For large PDFS: (c) average squared magnitude of the transfer function and PDL, (d) DGD versus wavelength.

To get an idea of the PDL and DGD to be expected in this device we tune our prototype to minimize PDFS as was done for measuring the data in Fig. 3.4(c). We record the average squared magnitude of the transfer function for all possible polarizations, find the PDL and detect the DGD as a function of wavelength, Fig. 3.5(a) and (b) (*FSR* ≈ 42.7 GHz). The peaks in PDL and DGD are due to the different transfer functions for the orthogonal polarizations. When the birefringence control voltage deviates from the optimum operating point, double peaks are observed in the PDL and DGD curves. We then introduce PDFS in a second experiment and consequently find strong PDL and DGD as plotted in Fig. 3.5(c) and (d). The squared magnitude of the transfer function is averaged over all possible polarizations, and therefore the measured ER is small. Thus, one could minimize PDFS by minimizing the PDL and DGD peaks at destructive interference, especially when at the single peaks in Fig. 3.5(a, b) (upper traces) the PDFS is minimum. The ER of the average squared magnitude of the transfer function for all possible polarizations is maximal when the PDFS is minimal.

So far we demonstrated a tunable DI with reduced PDFS. There remains the problem of the inaccurate step size and the hysteresis of the actuator, so we need means for a stable and precise setting of the time delay.

3.1.4 Time Delay Control

In practical application the absolute time delay needs to be controlled and mechanical instabilities need to be mitigated. To this end we design an active feedback control circuit. In Fig. 3.6(a), the setup of the time delay and phase control is illustrated. A pilot tone at frequency f_p counter-propagates with respect to the communication signal at f_c. Circulators are used for separation of the pilot and signal tone. Because f_p and f_c can be widely different, appropriate filters would also allow a co-propagating arrangement without circulators [68]. For each DI, the pilot tone is detected with a low bandwidth photodiode, sampled for processing in the digital domain and fed to the respective control circuit.

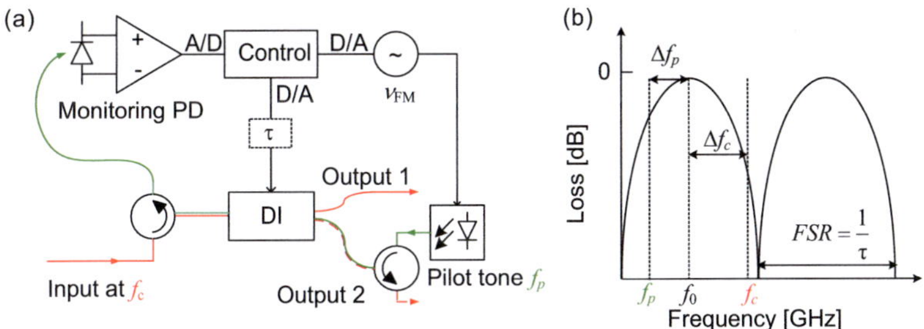

Fig. 3.6: Schematic of control circuit. (a) Time delay and phase control setup of DI, (b) ideal power transfer function between DI "Input" and "Output 2".

To set the absolute time delay accurately one starts off from a known time delay and then counts the number of constructive and destructive interference fringes m_{max} and m_{min}, respectively, when adjusting the delay. The associated change $\delta\tau$ of the time delay τ can be derived by counting the number of minima and maxima that a signal undergoes when detuning the delay

$$|\delta\tau| = \frac{m_{\max} + m_{\min}}{2 f_p}. \tag{3.1.8}$$

Since the direction of movement is known, we can estimate the time delay τ, and $\tau\times c$ (c is the speed of light) is expressed as multiple of the pilot tone wavelength. The remaining delay in distance is less than the wavelength of the pilot tone, which effectively introduces a phase offset to the signal. The remaining phase delay can be found by measuring the power swing of the pilot tone while scanning the delay, and by evaluating the pilot tone power deviation from the average power value at the end of the actuator travel.

For measuring the remaining phase delay more precisely [C19] dithering the DI delay may be used [68][69]. However, a mechanical DI dither introduces a deterioration of the transmitted signal data quality. So we decided for dithering the pilot tone frequency f_p harmonically according to $f(t) = f_p + \delta v_d \sin(2\pi v_{FM} t)$ using a frequency deviation δv_d with a modulation frequency v_{FM}. Fig. 3.6(b) shows the power transfer function and denotes its FSR, the DI's operating frequency f_0, and the signal carrier frequency f_c. Assuming that the pilot tone is connected to output 2 with f_0 being the frequency with constructive interference in the output, the frequency offset between pilot tone and DI is denoted $\Delta f_p = f_p - f_0$.

The phase monitoring photodiode in Fig. 3.6(a) produces an output current

$$\begin{aligned}
I_{ac}(t) \propto &-J_1\left(2\pi\delta v_d\tau\right)\sin\left(2\pi\Delta f_p\tau\right)\sin\left(2\pi v_{FM}t\right) \\
&+ J_2\left(2\pi\delta v_d\tau\right)\cos\left(2\pi\Delta f_p\tau\right)\cos\left(4\pi v_{FM}t\right) + ...
\end{aligned} \tag{3.1.9}$$

The symbols $J_{1,2}$ stand for the Bessel functions of order 1 and 2. The amplitudes of $\sin(2\pi v_{FM} t)$ and $\cos(4\pi v_{FM} t)$ can be extracted with a numerical lock-in scheme. τ is known with the fringe-counting measurement. Therefore the terms $J_{1,2}(2\pi\delta v_d \tau)$ can be estimated with good accuracy, and we calculate Δf_p from the remaining trigonometric terms $\sin(2\pi\Delta f_p \tau)$ and $\cos(2\pi\Delta f_p \tau)$. With Δf_p and pilot tone f_p known, the frequency f_0 of the transfer function's maximum is found, which has an offset Δf_c from the carrier frequency f_c. Therefore, based on the known offset Δf_c, a control loop can be set up to fine-tune the delay τ for the wanted offset Δf_c, which fixes the phase difference in both arms of the DI. Using the same pilot tone, two DIs in an IQ demodulator can be locked with a defined relative phase offset, i. e., $\pi/2$.

3.1.5 Measurement and Experiment Result

To demonstrate the quality of the tunable delay, we perform experiments with a long a medium and a short delay at 100 ps, 35.7 ps, and 10 ps. The average spectral responses over all possible polarizations, together with PDL versus wavelength are shown in Fig. 3.7. The plots show from left to right FSRs of 10 GHz (resolution limited), 28 GHz, and 100 GHz. No matter what the delay is the plots show an average ER of more than 20 dB with a minimum PDFS at the two outputs. The PDL around the frequencies of constructive interference are close to zero for all delays.

We then tested the performance across the spectral range. We performed three measurements with a *FSR* = 42.7 GHz (delay of 23.4 ps) at different wavelengths, Fig. 3.8. Similar performances have obtained in each case.

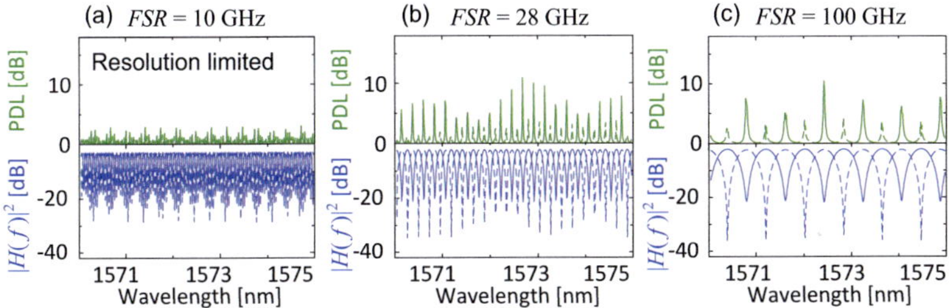

Fig. 3.7. Average spectral responses (blue) of the two outputs (solid line for output 1 and dashed line for output 2) over all possible polarizations and PDL (green) at free spectral range, *FSR* = 10 GHz (a), 28 GHz (b), and 100 GHz (c).

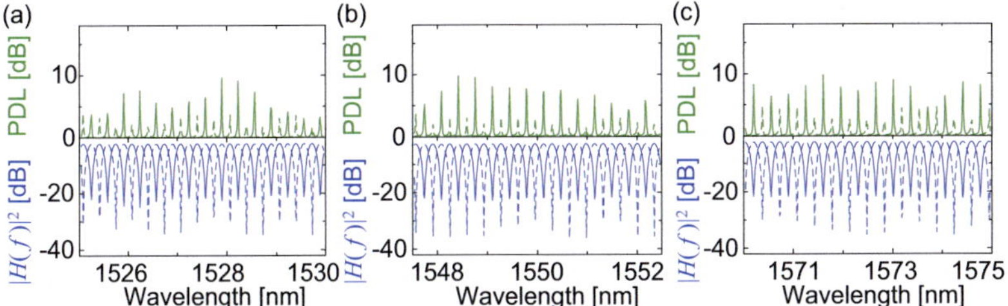

Fig. 3.8. Average spectral responses (blue) of the two outputs (solid line for output 1 and dashed line for output 2) over all possible polarizations and PDL at wavelength 1525 nm – 1530 nm (a), 1545 nm – 1555 nm (b), and 1570 nm – 1575 nm (c) with *FSR* = 42.7 GHz.

The accuracy of the absolute delay control algorithms has been tested next. We first set the delay interferometer at 700 GHz FSR. We then used a pilot tone at 1550.12 nm to perform the fringe count. The DI is tuned from 700 GHz to 600 GHz and down till 20 GHz, then up to 800 GHz and finally back to 700 GHz. The target value of each of the 40 measurements has been cross-examined with an optical spectrum analyzer (OSA) by connecting the DI input to an ASE source. Fig. 3.9(a) shows the relative deviation between the set FSR and the respective OSA measurement. The maximum deviation is ±0.08 %, which is mostly limited by the repeatability of the OSA measurement.

We then were interested in the accuracy of setting the DI to a particular phase. For this we modulated the pilot tone with ν_{FM} = 60 Hz and $\delta\nu_d$ = 2 GHz and used to scan the phase offset over 360° at *FSR* = 40 GHz. Results at *FSR* = 80 GHz are also shown. In Fig. 3.9(b), the deviations between the set phase offset and the cross-examinations with respect to an OSA are shown. A maximum error of ~2° is found. A 6 hours phase deviation measurement is performed with the pilot tone being modulated with ν_{FM} = 30 Hz and $\delta\nu_d$ = 750 MHz. A phase deviation < 2° and a standard deviation of 0.3° have been measured, see Fig. 3.9(c).

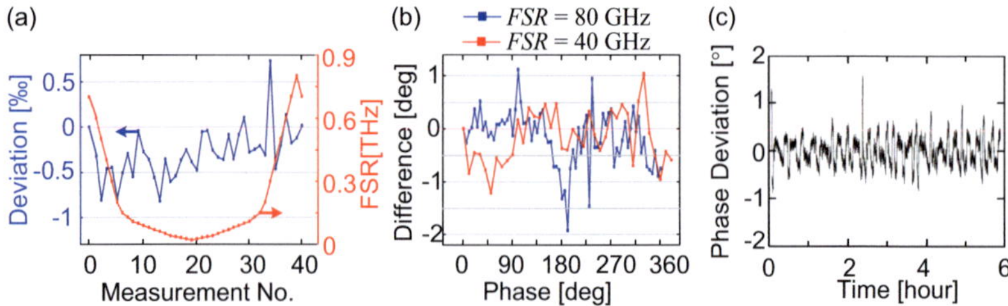

Fig. 3.9. The plots (a) the accuracy for setting the absolute time delay (blue curve) when setting the FSR to particular value (red curve) for a measurement cycle between 800 GHz and 20 GHz, (b) the accuracy for setting a particular two delays with $FSR = 40$ GHz —•— red, $FSR = 80$ GHz —•— blue for a measurement cycle from 0° to 360° and (c) absolute deviation from set value over time when using the stabilization setup.

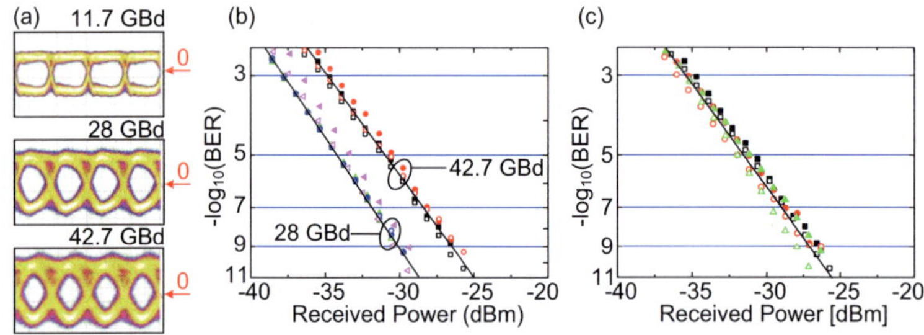

Fig. 3.10. Measurement results of the proposed DI at various bit rates and comparison with a commercial DI (a) eye diagrams of NRZ-DQPSK I channel at 11.7 GBd, 28 GBd, and 42.7 GBd, (b) BER versus received power at 28 GBd (▲△ green, for I and Q channels at polarization 1, •○ blue for I and Q channels at polarization 2), and 42.7 GBd (■□ black for I and Q channels at polarization 1, •○, red for I and Q channels at polarization 2) for orthogonal polarizations of the tunable DI and single polarization of the typical DI (◄◁ magenta for I and Q channels), and (c) BER versus received power at 42.7 GBd at different wavelengths (■□ black for I and Q channels at 1545.56 nm, •○ red at 1550.12 nm, and ▲△ green at 1560.61 nm).

The outputs of the DI were then connected to a 50 GHz balanced detector. The balanced receiver was built with a conventional photodiode combined with an inverted photodiode through a RF combiner. Demodulation of NRZ-DQPSK signals was performed at 11.7 GBd, 28 GBd and 42.7 GBd with a DI delay of 1 symbol duration. Eye diagrams of the inphase channel captured with a DCA are shown in Fig. 3.10(a). BER measurements were also performed with a PRBS sequence of $2^7 - 1$ at symbol-rates of 28 GBd and 42.7 GBd as depicted in Fig. 3.10(b). For the two orthogonal polarizations the required received power for fixed BER is almost equal for both symbol-rates with the proposed DI. The performance is comparable with a commercial available DI in the same measurement setup. In Fig. 3.10(c), similar received power requirement for fixed BER is observed over a broad wavelength range.

3.1.6 Polarization Division Differential Direct Detection Receiver

The tunable delay interferometer can be further extended into a polarization division differential direct detector. We show in a step-by-step manner that the four DIs normally used for a polarization diversity scheme with Inphase (I) and Quadrature (Q) phase detection can be simplified into single DI configuration.

The general concept is first explained on a conventional polarization division differential direct detector as shown in Fig. 3.11(a). Its optical frontend includes 4 DIs which have a time delay τ that is optimized with respect to the symbol rate of the detected signal. The signal is first split into the two polarization components E_s and E_p (which in general do not correspond to the transmitter's polarization components) by a polarization beam splitter (PBS). Then each polarization is fed into 2 orthogonal DIs ($\pi/2$ relative phase offset) (I- and Q-DIs) and four balanced detectors are used to detect the signal components.

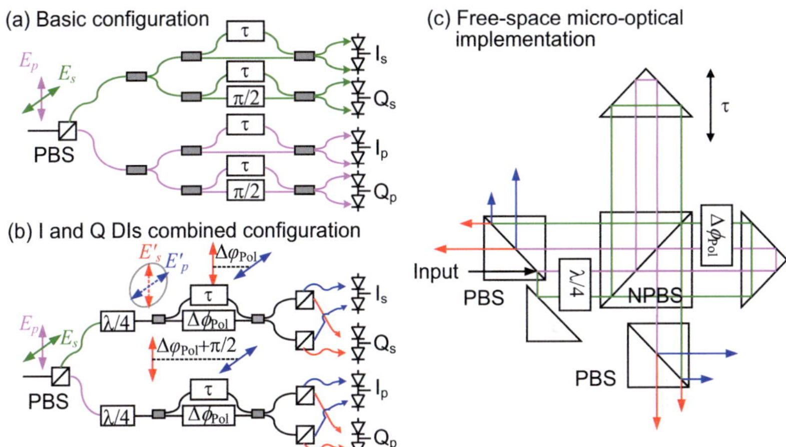

Fig. 3.11. Schematics of polarization diversity self-coherent receivers (a) conventional configuration with 4 DIs, (b) I and Q DIs combined configuration with 2 DIs, (c) free-space micro-optical implementation where all elements are folded into 1 DI only.

The DI as described above can be simplified to combine the I- and Q-DIs in a single DI. This novel concept is shown in Fig. 3.11(b). The scheme works as follows. First a PBS splits off the two polarizations E_s and E_p. Next we would like to determine the I- and Q-phase component of each of the polarizations. So we use quarter waveplates (QWP) to convert the E_s and E_p components into circular polarizations (which provide us for each path a new pair of s and p fields E'_s and E'_p. Please note that the real advantage of this scheme is that we have E'_s and E'_p fields of equal power!). Each circular polarization is then fed into a separate DI. One branch of the DI comprises the time delay and the other branch a birefringent element (e.g., a liquid crystal $\Delta\phi_{Pol}$). The delay may then be set for the E'_p-component to provide the I-phase component. The birefringent element may then be used to set the phase difference in the short branch to $\Delta\varphi_{pol} + \pi/2$ resulting in a $\pi/2$ relative phase offset for the E'_s-component with respect to the E'_p-component. The E'_s-component will thus then provide us the Q-phase term. Using PBSs at the DI outputs the E'_p and E'_s fields can easily be separated and combined into the balanced detectors. Thus we have reduced the number of required DIs to only 2 DIs in this newly proposed configuration.

In Fig. 3.11(c), we propose an additional reduction of required components in a free space optics implementation. Using a mirror, the two orthogonal polarizations are mapped into the same DI configuration with two beams propagating in parallel sharing the same optical elements. The signals are split and combined with one single NPBS and reflected back by two corner reflectors. One reflector is mounted on a movable actuator that introduces a time delay τ. In the other branch, the birefringent element is set to align the orthogonal relative phase offset between the I and Q-components. Two PBSs (including the one at the input) are used to separate the signals into p and s polarizations. The signals can be coupled into fibers or directly to photodiodes providing electrical signals for further processing. With this novel setup, four logic DIs in a polarization diversity self-coherent detection scheme are folded into one single and compact Michelson delay interferometer structure with one single actuator tuning the time delay.

3.1.7 Conclusion

A tunable free-space optical delay interferometer has been modeled and presented. Its delay is controlled with a pilot tone. The maximal time delay is up to 100 ps, which allows a reception of symbol rates > 10GBd. Its PDFS is mitigated by utilizing an adjustable birefringent liquid crystal. A more complex but very compact design for polarization multiplexed D(m)PSK signal detection is also presented.

3.2 Four-in-one Interferometer for Coherent and Self-coherent Detection

The content of this section is a direct copy of the Journal publication [J1]:
J. Li, M. R. Billah, P. C. Schindler, M. Lauermann, S. Schuele, S. Hengsbach, U. Hollenbach, J. Mohr, C. Koos, W. Freude, and J. Leuthold, "Four-in-one interferometer for coherent and self-coherent," Opt. Express, submitted.
Part of the content of this section has been filed in a patent [P1]:
J. Li, M. Lauermann, S. Schuele, J. Leuthold, and W. Freude, "Optical detector for detecting optical signal beams, method to detect optical signals, and use of an optical detector to detect optical signals," US patent 20, 120, 224, 184, 2012.
Minor changes have been done to adjust the notations of variables, color, line type and position of the figures.

A compact micro-optical interferometer is presented that combines two optical 90° hybrids or, alternatively, four delay interferometers into one interferometer structure sharing one tunable delay line. The interferometer can function as a frontend of either a coherent receiver or of a self-coherent receiver by adjusting the waveplates and the delay line. We built a prototype on a LIGA bench. We characterized the device and demonstrated its functionality by successful reception of a 112 Gbit/s signal.

3.2.1 Introduction

For coherent communication to become practical, cheap coherent and self-coherent receivers that require as few elements as possible are needed. Many research groups have been investigating solutions based on integrated optics [70][71][72]. Integrated optics solutions offer small dimensions and possible co-integration with photo-diodes and electronic circuits. Yet, the most common technique for commercial devices [73] is still based on free-space optics because of the stability and high quality of the optical characteristics. Another advantage of free space solutions is the large operation wavelength range (several 100 nm, e.g., from 1300 nm [74] to 1550 nm [J5]), and the option to continuously tune the time delay which is useful for self-coherent reception [J5].

In contrast to integrated optics where many elements can be co-integrated, free-space optics is limited in the number of physical components due to space and cost restrictions. A polarization and phase diverse coherent receiver requires 2 optical 90° hybrids. Each optical 90° hybrid comprises several beam splitters, phase shifters and reflectors for beam alignment, see Fig. 3.12(a). A self-coherent receiver consists of either a delay line plus 2 optical 90° hybrids or 4 delay interferometers (DI), see Fig. 3.12(f) and (g). The latter requires 3 additional time delay arms in comparison to the former. Each DI needs to be controlled individually in order to match the signal's symbol rate $(1 / T_s)$, and the phase of each DI needs to be adjusted to guarantee orthogonality between the inphase (I) and quadrature (Q) DIs.

Therefore, efforts to reduce the number of hybrids/DIs are necessary. One method is to use polarization insensitive hybrids/DIs and separate the polarizations at the outputs [33]. However, polarization-insensitive hybrids/DIs are intricate to fabricate. Additionally, because the polarizations are separated at the outputs, the scheme is very sensitive to unwanted polarization rotation or birefringence before the polarizations are separated. Any spurious polarization rotation or birefringence would directly translate into coherent polarization crosstalk. The second method is to spatially separate the two polarizations at the input and lead them in parallel into the same optical components, which can be easily assembled with free-space optics. However, this method requires highly accurate optical alignment.

Being a high precision X-ray lithography technology, LIGA technology (lithography, galvanic electroplating, molding) [75] can be used for fabricating passive alignment structures for optical elements with tolerances down to 200 nm and maximum heights up to 1 mm. By snapping micro-optical elements into the LIGA alignment structure, one can reduce the dimensions of the setup [C15]. In addition, good optical performance is guaranteed by the mature coating technology of free-space optical elements.

In this section we introduce a compact micro-optical interferometer in LIGA technology. The number of optical components used in the interferometer is minimized by reusing the optical interfaces up to four times. Our scheme folds two optical $90°$ hybrids for a coherent receiver and four delay interferometers for a self-coherent receiver into a single interferometer structure with one single time delay adjustment. It can be switched from a polarization diverse coherent receiver to a polarization diverse self-coherent receiver by simply adjusting the birefringent waveplates.

3.2.2 Interferometer Principle and Structure

To better explain the path towards a component reduction for coherent and self-coherent receivers we begin with a description of the basic structures.

In a conventional optical polarization and phase diverse coherent receiver, the incoming signal (solid line in Fig. 3.12(a)) is split by a polarization beam splitter (PBS) in two orthogonal polarizations, namely s (perpendicular to the incident plane, green) and p (parallel with the incident plane, magenta) polarization, each of which is then superimposed to a local oscillator (LO, dashed line) at the same polarization in an optical $90°$ hybrid. At the outputs, the inphase (I) and quadrature phase (Q) components of the sum and difference of two beams are received by pairs of balanced detectors as shown in Fig. 3.12(a).

In a self-coherent receiver, instead of the local oscillator, a delayed copy of the signal is combined with itself in an optical $90°$ hybrid (Fig. 3.12(f)), or in two delay interferometers with orthogonal phase offset (Fig. 3.12(g)).

Therefore a polarization and phase diverse coherent receiver requires 2 PBS, 8 couplers, and 2 phase shifters. A polarization and phase diverse self-coherent receiver requires 2 or 1 PBS, 8 or 10 couplers, 1 or 4 tunable delay lines and 2 phase shifters.

Here, we will show how to build a receiver frontend that only requires 2 PBS, 1 non-polarizing beam splitter (NPBS), 3 reflectors for folding beams to reuse optical surfaces,

3 waveplates, and 1 tunable delay line. Thus, we can reduce the total number of optical components from 12…17 (not taking into account the reflectors that typically are needed to align the beams with the components) to a number of only 7 optical components plus 3 reflectors. At the same time, this device can be switched between a coherent receiver and a self-coherent receiver frontend.

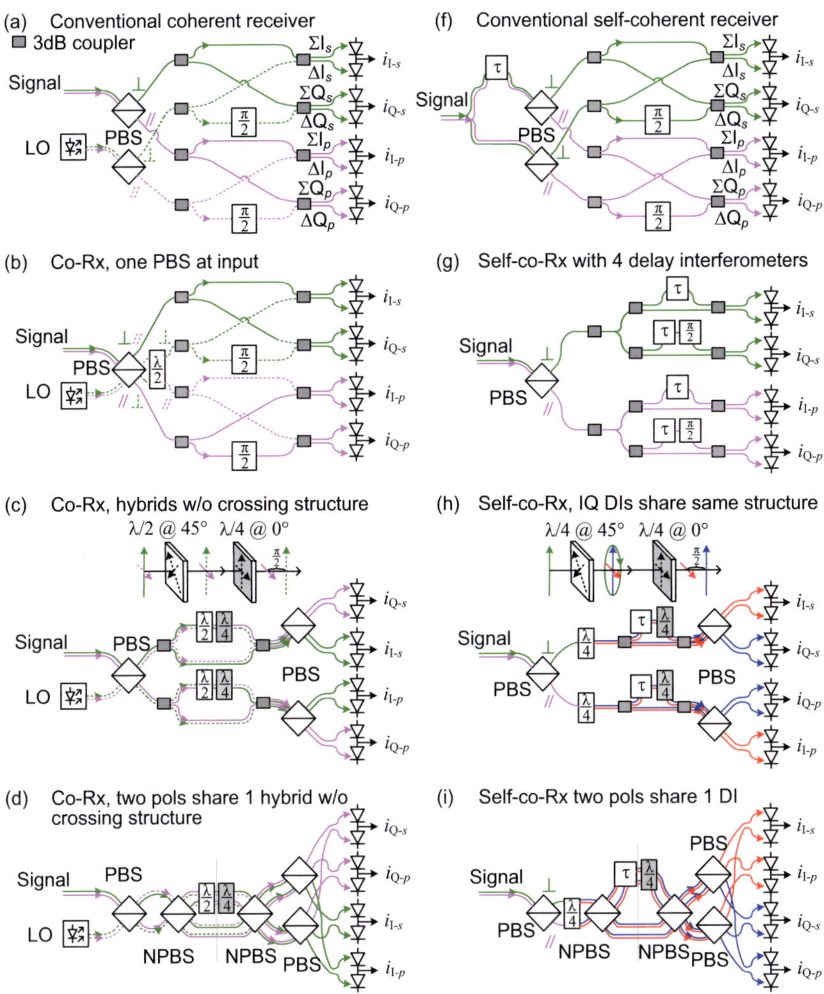

Fig. 3.12. Schematic illustration of a polarization and phase diverse coherent receiver and self-coherent receiver frontend. The left column presents from top to bottom the transformation from a conventional coherent receiver (a) to a free-space optics interferometer configuration (e). The right column (f) to (j) presents a similar transformation for the self-coherent receiver. (e) and (j) are on the next page. Both receiver types can be realized with the same device. $\perp$: *s*-polarized light; //: *p*-polarized light.

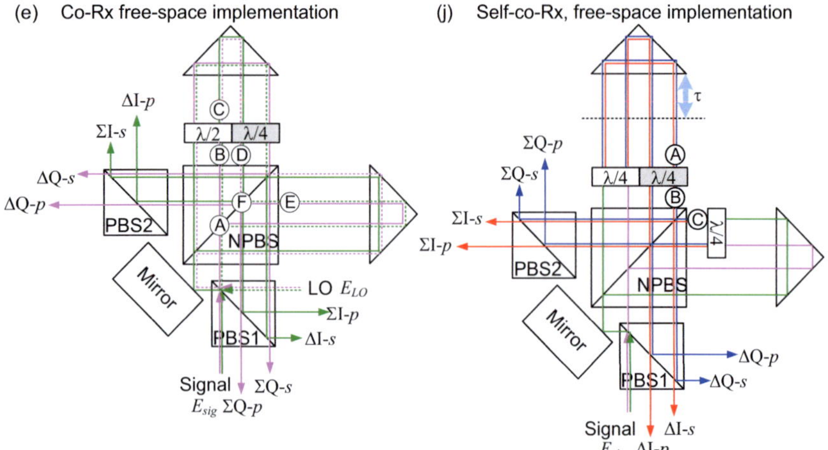

Fig. 3.12. Schematic illustration of a polarization and phase diverse coherent receiver and self-coherent receiver frontend. The left column presents from top to bottom the transformation from a conventional coherent receiver (a) to a free-space optics interferometer configuration (e). The right column (f) to (j) presents a similar transformation for the self-coherent receiver. (a), (b), (c), (d), (f), (g), (h), and (i) are on the previous page. Both receiver types can be realized with the same device. ⊥: *s*-polarized light; //: *p*-polarized light.

Coherent receiver. We first reduce the number of PBSs at the input from 2 to 1. In Fig. 3.12(b), the signal and the LO are led into the PBS from different sides. After the PBS, a half waveplate (HWP) at 45° rotation with respect to the PBS is used to flip the split polarization states of the LO to their orthogonal states. Then the signal and LO at the same polarizations are mixed in the optical 90° hybrids.

In free-space optics, the crossing structure in the middle of the optical 90° hybrid of Fig. 3.12(b) requires several reflectors. To further reduce the number of optical elements, the optical 90° hybrid can be built in a different way, Fig. 3.12(c). After the PBS, the signals are split in two couplers. At one arm of each coupler, a HWP at 45° rotation with respect to the PBSs is introduced to convert the signal and LO onto their orthogonal polarizations. Thus the signal can be superimposed with the LO at the same polarizations at the second coupler. A quarter-waveplate (QWP) at 0° is positioned before the second coupler to provide the necessary 90° phase shift for creating the quadrature signal. Now we have again two optical hybrids after the input PBS. Compared to Fig. 3.12(b), three more waveplates and two more PBSs are required while saving four couplers, 2 phase shifters, and avoiding crossings.

The two optical hybrids can then be built into one optical hybrid configuration by spatially overlapping them as in Fig. 3.12(d). Here the couplers are realized by non-polarizing beam splitters (NPBS). By folding the configuration along the gray line in the figure, we finally get the interferometer in Fig. 3.12(e), where only 1 NPBS and 2 PBSs are needed instead of 2 NPBS and 3 PBS in Fig. 3.12(d).

Self-coherent receiver. Starting from Fig. 3.12(g), we first reduce the number of DIs from 4 to 2 as in Fig. 3.12(h). After the PBS, a QWP at 45° rotation with respect to the PBSs is introduced at each polarization to convert the linearly polarized light into circularly polarized light. The circularly polarized light has new s (blue line) and p (red line) components with equal power. They share the same DI and are separated at the outputs by PBSs. By using a QWP at $0°$, $90°$ phase shift is introduced between the new s and p components, thus that one polarization forms an inphase component (e.g., the red line) while the other forms a quadrature phase component (e.g., the blue line).

The two delay interferometers can be combined within the same DI structure as in Fig. 3.12(i). Finally, the configuration can be re-arranged as a free-space optical interferometer with only one NPBS and two PBS as shown in Fig. 3.12(j). One may now compare the coherent receiver, Fig. 3.12(e) with the self-coherent receiver of Fig. 3.12(j). The only differences are in an exchange of waveplates and an additional waveplate, in a non-zero delay for the self-coherent receiver, and in a LO input for the coherent receiver. Thus both the configurations can be easily converted back and forth by using adjustable waveplates and a tunable delay. Tunable waveplates can be realized with liquid crystal cells, and the tunable time delay can be obtained by mounting the reflector onto a motor. A mathematic expression of the concept can be found in the Appendix A.2.

3.2.3 Free–space Coherent & Self–coherent Receiver Design and Implementation

The free-space scheme comprises a micro-optical bench that serves as a guiding structure for high-precision alignment of the optical elements. The fabrication steps are as follows:

The micro-optical bench is designed and modeled with a ray-tracing tool, and the resulting structure exported as a CAD layout. Then alignment constructs are added around the optical micro-elements. The complete 3D volume assembly is then reduced to a 2D outline, which serves as input data for the X-ray mask.

First, a premask is manufactured using e-beam lithography. A gold layer with a height of about 2 μm is structured on a 4 μm thick titanium foil, see Fig. 3.13(a). In the next step, the premask is used for the production of a master mask. The master mask is an invar (nickel iron alloy with low thermal expansion) foil carrying the structured gold absorber layer with a height of about 25 μm, which is fabricated with gold electroplating, Fig. 3.13(b). Both gold layers absorb X-rays, while the titanium foil or the invar foil is essentially transparent. A foil with PMMA resist having a thickness of 610 μm is glued to the silicon wafer. After deep X-ray lithography, a developing process leads to high aspect ratio PMMA alignment structures in the micro-optical bench, Fig. 3.13(c). For mass production, such structures are strengthened with a nickel electroplating process, and the resulting nickel molding tool can be used for hot embossing. The complete process is often described as LIGA technology [75] (*li*thography, *ga*lvanic electroplating, and molding).

Single benches with dimensions of 21×22 mm^2 are diced from the wafer by laser cutting using a pulsed NdYag laser. The dimension of the single bench may be further reduced if smaller optical elements were chosen instead of commercially available elements.

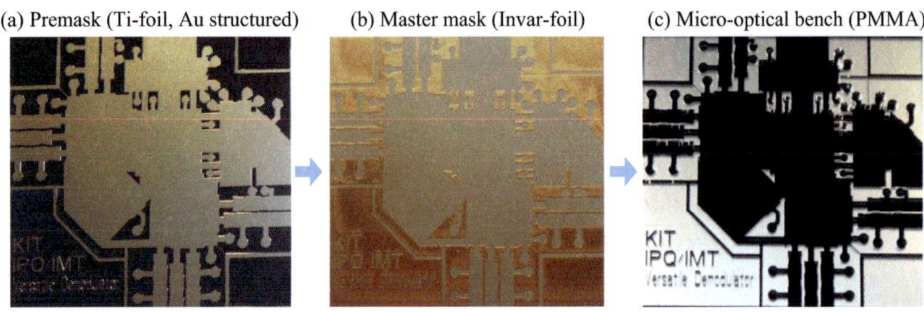

Fig. 3.13. Fabrication steps for the LIGA process. Pictures of the (a) premask (Ti-foil with Au structures), of the (b) master mask (invar-foil), and of the final (c) micro-optical bench (PMMA).

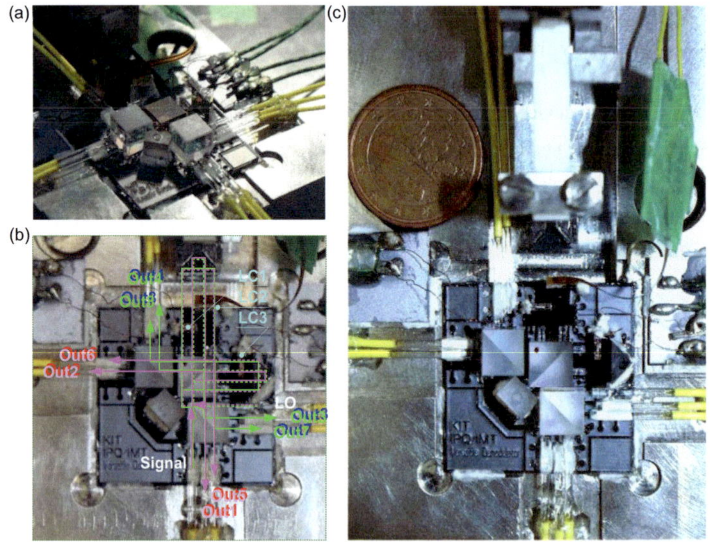

Fig. 3.14. Photograph of a complete micro-optical interferometer. (a) Side view; (b) Top view with beam paths; (c) Top view of the complete micro-optical interferometer including the piezo-motor and in comparison with one Euro cent.

A prototype of micro-optical interferometer has been fabricated by positioning the optical elements using the alignment and guiding features of the micro-optical bench, see Fig. 3.14. Grin lenses with a diameter of 1 mm are used to couple the optical beams from and to the fiber. PBS and NPBS are chosen to be $5 \times 5 \times 5$ mm^3 which is commonly available on the market. Porro prisms are coated with metallic coating on their orthogonal sides. One Porro prism is attached to a piezo-motor which acts as a tunable time delay and an adjustable phase shifter. A mirror cube is used to deflect the second polarization (*s* light) to the NPBS. Liquid crystal (LC) cells serve as waveplates whose birefringence can be changed by adjusting the applied voltage. Two LC cells have axes rotated with respect to the PBS eigenstates by $-45°$, and one has axes aligned with the PBS. The LC cells are positioned such that the beams reflected by the Porro prisms would only pass the waveplates once, as indicted in Fig. 3.12(e)(j) and Fig. 3.14(b). Finally, the optical elements are glued to the bench. The 8 outputs for the I and Q components and the two polarizations are named 'Out1' to 'Out8'. They correspond to different ports of the coherent and self-coherent receiver as listed in Table 3.1.

Output number	Out1	Out2	Out3	Out4	Out5	Out6	Out7	Out8
Coherent	ΣQ-p	ΔQ-p	ΣI-p	ΔI-p	ΣQ-s	ΔQ-s	ΔI-s	ΣI-s
Self-coherent	ΔI-p	ΣI-p	ΔQ-p	ΣQ-p	ΔI-s	ΣI-s	ΔQ-s	ΣQ-s

Table 3.1. Outputs numbering in coherent and in self-coherent receiver.

3.2.4 Characterization and Testing

The free space receiver frontend was tested for its performance in a self-coherent and in a coherent receiver.

We first tested the device as a self-coherent receiver. For this purpose, the LC cells were adjusted to act as QWPs at $-45°$ (LC1 and LC3), and as a QWP at $0°$ (LC2) as in our schematic of Fig. 3.12(j). The LC2 cell at $0°$ may also be used to compensate polarization dependent frequency shift (PDFS) within the setup [J5].

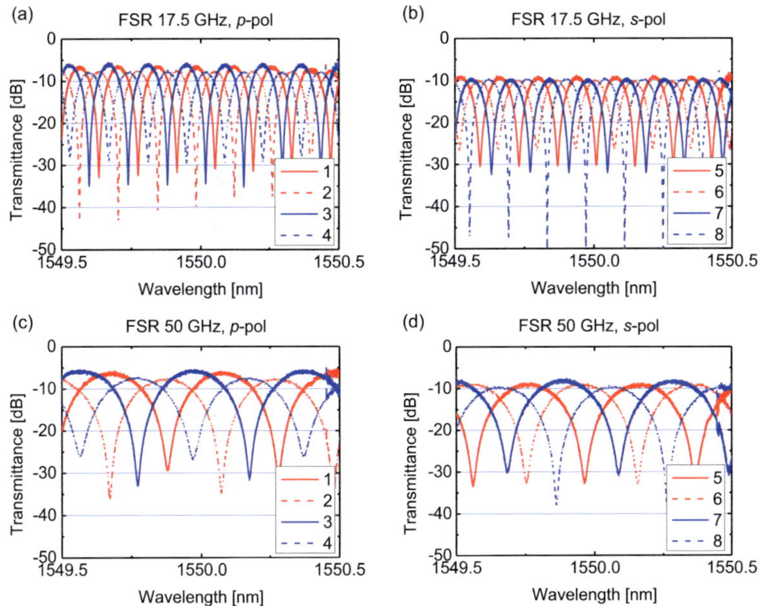

Fig. 3.15. Micro-optical interferometer as self-coherent receiver. Spectral responses of outputs 1 (red solid line), 2 (red dash line), 3 (blue solid line) and 4 (blue dash line) at p-polarization in a wavelength range 1549.5 nm … 1550.5 nm. Free spectral range (a) 17.5 GHz, and (c) 50 GHz. Spectral responses of outputs 5 (red solid line), 6 (red dash line), 7 (blue solid line) and 8 (blue dash line) at s-polarization for 1549.5 nm … 1550.5 nm. Free spectral ranges (b) 17.5 GHz and (d) 50 GHz.

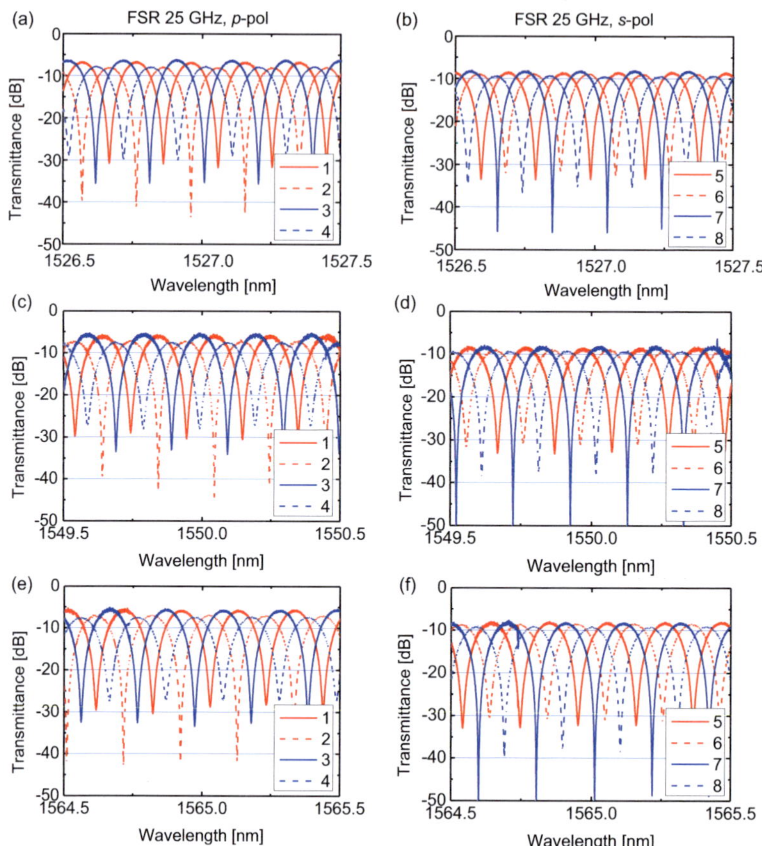

Fig. 3.16. Micro-optical interferometer as self-coherent receiver. Spectral responses at outputs 1 (red solid line), 2 (red dash line), 3 (blue solid line) and 4 (blue dash line) for *p*-polarization and FSR = 25 GHz. Wavelength range (a) 1526.5…1527.5 nm, (c) 1549.5 …1550.5 nm, (e) 1564.5…1565.5 nm. Spectral responses of outputs 5 (red solid line), 6 (red dash line), 7 (blue solid line) and 8 (blue dash line) at *s*-polarization and FSR = 25 GHz. Wavelength ranges (b) 1526.5…1527.5 nm and (d) 1549.5…1550.5 nm, (f) 1564.5…1565.5 nm.

The static performance of the micro-optical interferometer is characterized by wavelength-sweeping a C-band tunable laser with a wavelength resolution of 1.44 pm. The micro-optical interferometer is measured with the minimal and maximal FSR by adjusting the piezo-motor to its farthest and closest position with respect to the Porro prism. In Fig. 3.15 we plot the spectral response of outputs Out1…Out8 at *p* and *s* polarization adjusted to a FSR of 17.5 GHz and 50 GHz. The outputs are named as in Table 3.1. We see that the outputs represent constructive and destructive ports of the I and Q components. The extinction ratios for all the outputs are larger than 18 dB, which is sufficient for signal reception [50]. The excess insertion losses for all the outputs are below 4 dB. The insertion losses for output 5, 6, 7 and 8 are slightly higher than for output 1, 2, 3, and 4. This is because the *s* polarization is deflected by the mirror cube which introduces extra absorption and misalignments. We conclude that the device keeps good extinction ratio and insertion loss over a tuning range of 17.5...50 GHz, corresponding to a delay tuning range of 20...57 ps. The device has potential to cover a larger time delay range, e. g., down to zero delay, by reducing the tunable path when dicing the chip.

The spectral response for various free spectral ranges (FSR) is measured in the C band. In Fig. 3.16, we present the spectral response for FSR = 25 GHz near 1527 nm, 1550 nm and 1565 nm at p and s-polarization. The device has similar performance over the full C band.

We then tested the micro-optical interferometer as a coherent receiver. One LC cell which was positioned previously at −45° (LC3) is now set to "neutral" so that it does not change the polarization states of the beams. The other LC cell at −45° (LC1) is set as a HWP so that it flips the linear polarization states of signal and LO to their orthogonal states. Polarization extinction ratios are measured by comparing the power of p and s-light at the outputs with either the signal input only, or with the LO input only. The polarization extinction ratios for all outputs are listed in Table 3.2. The minimal polarization ER is 17.9 dB. The insertion losses for signal and LO at each output are also compared. The maximal difference is to be seen at Out4, where a 2.95 dB power difference is measured. There is non-zero path difference between the upper and the right paths in Fig. 3.14(b) due to technological constrains of our present design. These differences can be compensated by choosing proper lengths of optical patch cords, or by using digital signal processing.

Outputs	Out1	Out2	Out3	Out4	Out5	Out6	Out7	Out8
Sig. Pol. ER [dB]	19.4	19.2	22	20.7	18.4	17.9	20	19.6
LO Pol. ER [dB]	19.5	18	25.3	23	22.2	19.6	20.9	18.2
Insertion loss difference [dB]	0.03	1.29	1.97	2.95	0.83	0.88	0.26	1.07

Table 3.2. Polarization extinction ratios for the outputs of micro-optical interferometer as coherent detection scheme.

Finally, we tested the micro-optical interferometer with real data by receiving a 28 GBd PolMUX QPSK signal. The experimental setup is depicted in Fig. 3.17(a). A PolMUX QPSK signal is connected directly to the micro-optical interferometer via an erbium-doped fiber amplifier (EDFA) stage and an optical filter. The input polarization states at the input of the receiver are aligned with the polarization states at the transmitter so that no digital signal processing is required to demultiplex the polarizations. The signals are then received with four pairs of balanced detectors, and then sent to two real-time sampling scopes (Agilent optical modulation analyzer having a sampling rate of 80 GSa/s and a bandwidth of 32 GHz). No digital equalizer was applied. First, the micro-optical interferometer is configured as a self-coherent receiver. The constellation diagrams of the received signal in both polarizations are presented in Fig. 3.17(b). The signal quality is quantified with error vector magnitude (EVM) measurements. A value of EVM = 15 % is found, which is comparable with the signal quality for commercial devices. Second, the micro-optical interferometer is configured as a coherent receiver. The associated constellation diagrams are presented in Fig. 3.17(c), and the measured EVM are about 16%. The slightly worse performance of the coherent receiver is mostly due to the finite polarization extinction ratio and the insertion loss difference as shown in Table 3.2. This can be improved by using better-quality waveplates.

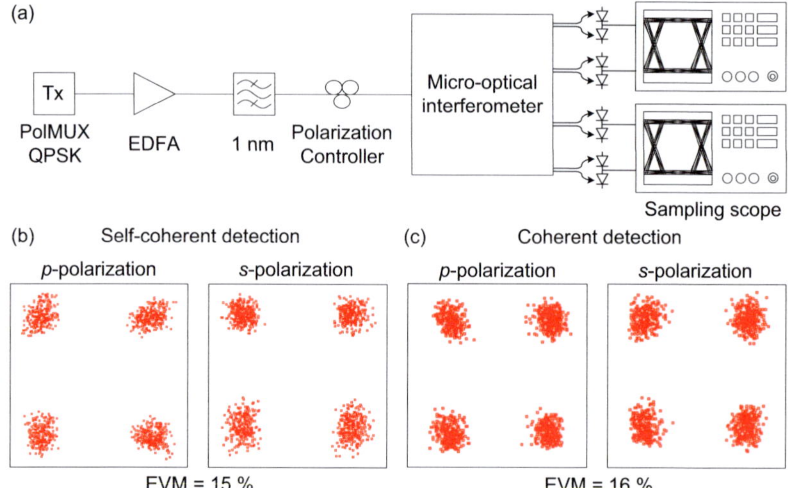

Fig. 3.17. Experimental setup and results for self-coherent and coherent receivers with the micro-optical interferometer serving as a frontend. (a) Schematic of the experiment setup. (b) Constellation diagrams of the signals received by a self-coherent receiver. (c) Constellation diagrams of the signals received by a coherent receiver.

3.2.5 Conclusion

We presented a micro-optical interferometer that can be switched between a self-coherent receiver and a coherent receiver. It folds four delay interferometers and two optical 90° hybrids into one single interferometer structure. With the help of a LIGA fabricated micro-optical bench, the four-in-one interferometer can be implemented on a 21×22 mm^2 chip with commercial available components. We measured excellent optical performance. Its function as a self-coherent and a coherent receiver frontend has been validated experimentally with a 112 Gbit/s PolMUX QPSK signal.

4 Self–coherent 100 Gbit/s Systems

Self-coherent receiver combines conventional differential direct detection with advanced digital signal processing. Compared to coherent receiver, it has the advantage of not using a local oscillator. It can be used to detect coherent signal including DBPSK, DQPSK, and 16QAM and polarization multiplexed signal [31][32][33][76][77]. Among these modulation formats, PolMUX-DQPSK at a symbol rate of 28 GBd has been considered as the most promising candidate for the next 100 Gbit/s system.

In this chapter, we will first present a self-coherent receiver comprising tunable delay interferometers, which allow the self-coherent receiver detecting signals at various symbol rates and different optical sampling rates. The principle of the self-coherent receiver will be explained and tested with a 28 GBd PolMUX-DQPSK signal by aligning the polarization with a polarization controller. This has been published in a conference paper [C21]. Then we will analyze the challenges for reception of a signal with arbitrary state of polarization and propose a DSP algorithm to relax the requirement of the polarization state at the input of the receiver. This has been published in a conference paper [C6]. Finally a self-coherent system using a variant of the PolMUX-DQPSK format is introduced, whereby the constellations in both polarizations are rotated by 45° relative to each other. A training sequence is used to estimate the fiber channel and a decision feedback circuit is applied to remove accumulated noise. PolMUX-DQPSK signal of 100 Gbit/s at arbitrary polarization state can be received. This part of the thesis has been published in Optics Express [J2].

4.1 Self–coherent Receiver with Tunable Delay Line

Part of the content of this section has been published in [C21]:
J. Li, K. Worms, P. Vorreau, D. Hillerkuss, A. Ludwig, R. Maestle, S. Schüle, U. Hollenbach, J. Mohr, W. Freude, and J. Leuthold, "Optical vector signal analyzer based on differential direct detection," in Proc. of 22nd Annual Meeting of the IEEE Photonics Society (LEOS 2009), Belek-Antalya (Turkey), paper TuA4, 2009 (Best Student Paper Award).

Self-coherent receivers usually only work with a fixed delay line of the built-in delay interferometers. Therefore the signal can be only detected at a certain symbol rate with a limited temporal resolution. In this section, we demonstrate a self-coherent receiver based on tunable delay interferometers (DI). Since the time delay in DI can be tuned between 0 ps and 100 ps, it can be adjusted to the delay that provides the optimum performance for a given symbol rate. We demonstrate the reception and analysis of several signal formats at different symbol rates. It includes a 112 Gbit/s (i.e., 28 GBd) PolMUX-DQPSK signal, 28 Gbit/s NRZ-OOK, 10 Gbit/s CSRZ (carrier suppressed return to zero), 28 Gbit/s DBPSK, and 42.7 GBd DQPSK signal.

4.1.1 Operation Principle

The schematic setup of a self-coherent receiver for receiving PolMUX-DQPSK signals is depicted in Fig. 4.1(a). The input signal is aligned to the eigen-states of the polarization beam-splitter (PBS) and then split in two orthogonally polarized beams. It is then processed by two pairs of tunable optical delay interferometers having (orthogonal) cosine and sine characteristics. The DIs are connected to balanced receivers. The optical input signal denoted by its complex notation $E_{x,y}(t) = A_{x,y}(t)\exp[j(\omega_0 t + \varphi_{x,y}(t))]$ at each polarization is split in two paths, and sent to two DIs. In the DI, the signal is further split in two arms having a delay difference τ. Delayed and non-delayed signals interfere at the output ports. Orthogonal interferometer characteristics are achieved by a relative phase offset in the arms of one DI with respect to the other DI amounting to $\phi_I = \pi/4$ and $\phi_Q = -\pi/4$, respectively. The balanced receivers' output currents of the inphase (I) and quadrature phase (Q) DIs are

$$I_{I_{x,y}} \propto \left|E_{x,y}(t)\right|\left|E_{x,y}(t-\tau)\right|\cos(\Delta\varphi_{x,y}(t)+\frac{\pi}{4}),$$

$$I_{Q_{x,y}} \propto \left|E_{x,y}(t)\right|\left|E_{x,y}(t-\tau)\right|\sin(\Delta\varphi_{x,y}(t)+\frac{\pi}{4}),$$

$$(4.1.1)$$

where $\Delta\varphi_{x,y}(t) = \varphi_{x,y}(t) - \varphi_{x,y}(t-\tau)$. These output currents then are electrically sampled with an Agilent sampling oscilloscope. An off-line software tool then extracts the complex current,

$$u_{x,y} = I_{I_{x,y}} + jI_{Q_{x,y}}.$$

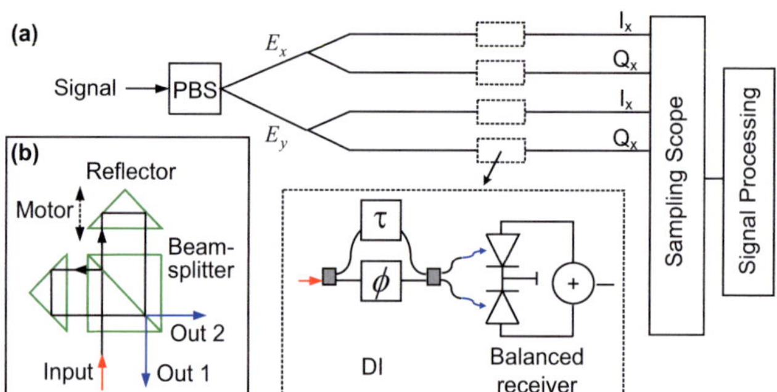

Fig. 4.1. (a) Self-coherent receiver scheme. (b) Free-space optics layout of tunable optical delay interferometer (DI) as explained in Section 3.1.

An approximation for the amplitude $|E_{x,y}(t)|$ and an exact relation for the phase $\varphi_{x,y}(t)$ at time $t = t_0 + n\tau$ (with respect to an arbitrary reference phase at t_0) leads to [33].

$$E_{x,y}(t_0 + n\tau) = \left|E_{x,y}(t_0 + n\tau) \cdot E_{x,y}(t_0 + n\tau + \tau)\right|^{\frac{1}{4}}$$

$$\times \exp\left(j\sum_{m=1}^{n}\left(\arg\left(u_{x,y}(t_0 + m\tau)\right) - \frac{\pi}{4}\right)\right)\exp\left(j\varphi_{x,y}(t_0)\right).$$

$$(4.1.2)$$

As the time delay τ in the delay interferometers can be tuned continuously, the reconstructed discrete field can be resolved with different temporal resolution.

4.1.2 Experimental Demonstrations

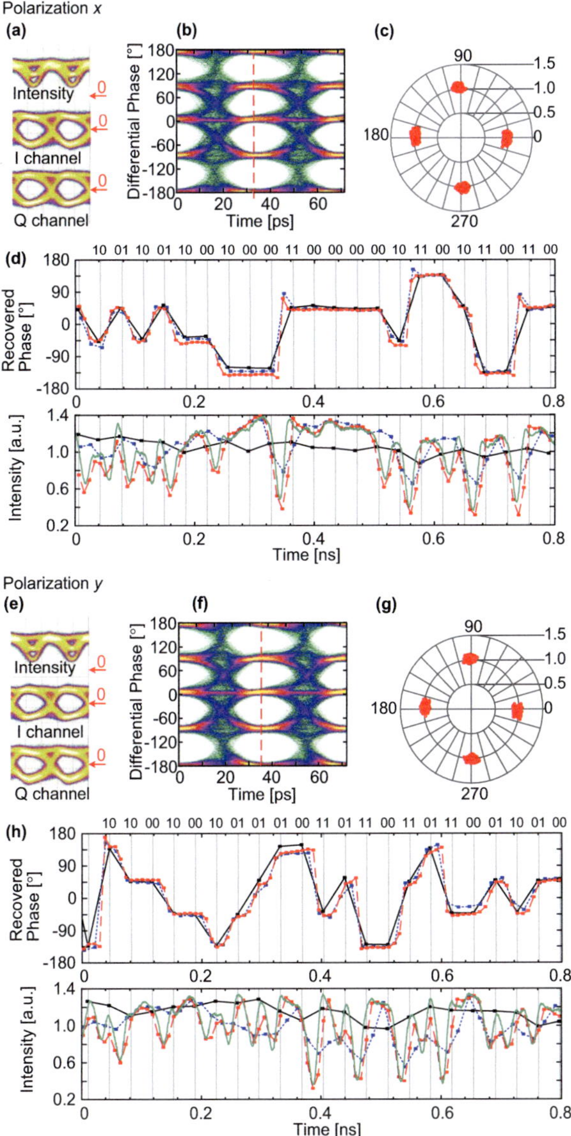

Fig. 4.2. Detection of 2,085 symbols. (a) & (e): Measured intensity and I&Q channel eye diagrams; (b) & (f): differential phase eye diagrams with sampled once per symbol ($SpS = 1$); (c) & (g): differential phase constellation diagrams of DQPSK signal for each of two orthogonal polarizations; (d) & (h): recovered intensity and phase as a function of time with $SpS = 1$ (– ▪ –), $SpS = 2$ (⋯ ▪ ⋯) and 4 (--- ▪ ---); Directly measured intensity (——) shown for comparison. Symbol frame centers marked in gray (|).

An optical 112 Gbit/s PolMUX-DQPSK signal with a symbol rate of 28 GBd at $\lambda = 1{,}555.75$ nm and a pseudo-random sequence length of $2^{11}-1$ is launched. The delay of the built-in delay interferometers is set to 35.7 ps corresponding to one symbol duration of 28 GBd. The I and Q waveforms are then detected with photodiodes (cut-off frequency of 75 GHz) and sampled by an Agilent sampling oscilloscope, Fig. 4.2(a)(e). The orthogonal polarizations were pre-aligned with a polarization controller. The recorded waveforms are subsequently off-line processed.

The differential phase, the recovered phase and the intensity can be calculated with Eq. (4.1.1) and Eq. (4.1.2). By overlaying the recorded 2,085 symbols, the differential-phase eye diagrams can be plotted, Fig. 4.2(b)(f) for the two polarizations, respectively. The constellation diagrams of the PolMUX-DQPSK are plotted in Fig. 4.2(c)(g) with differential phases and recovered amplitudes. The fully recovered phase and intensity of the signal are plotted in Fig. 4.2(d)(h) $(-\bullet-)$. By tuning the delay to 17.85 ps and 8.9 ps, i.e., to $SpS = 2$ $(\cdots\bullet\cdots)$ and $SpS = 4$ $(---\bullet---)$, it provides more accurate results. The recovered phase at $SpS = 4$ reveals the best temporal resolution, and the recovered intensity at $SpS = 4$ matches best with the measured intensity ($——$) that has been recorded as a check. The temporal resolution is limited by the bandwidth of the photodiodes.

Other modulation formats at various bitrates, namely a DQPSK signal at 42.7 GBd, a DBPSK signal at 28 Gbit/s, an NRZ-OOK signal at 28 Gbit/s, and a CSRZ signal at 10 Gbit/s are also tested with the self-coherent receiver (Fig. 4.3).

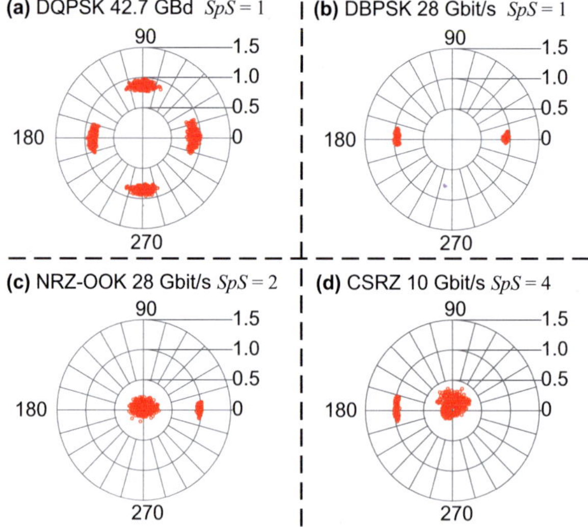

Fig. 4.3. Differential phase constellations of (a) DQPSK 42.7 GBd $SpS = 1$, (b) DBPSK 28 Gbit/s $SpS = 1$, (c) NRZ-OOK 28 Gbit/s $SpS = 2$, and (d) CSRZ 10 Gbit/s $SpS = 4$, with the samples positioned at the symbol center.

4.1.3 Conclusion

We presented a self-coherent receiver with tunable delay interferometers built in. It can be used to receive and analyze various modulation formats at different bit rates. The temporal resolution of the received signal can be tuned by adjusting the delay line of the delay interferometer. It has been shown that self-coherent receiver with smaller delay line performs better field reconstruction. However, the smallest temporal resolution is limited by the bandwidth and sampling rate of the ADCs at the receiver. When detecting PolMUX signal, the signal at the receiver needs to be aligned to the eigen-state of the receiver with the help of a polarization controller. This introduces extra hardware which has to be accurate and endless controlled at the speed of the polarization variation of the channel. In the next section, a receiver algorithm is developed for relaxing the requirement of the polarization controller.

4.2 Self–coherent Receiver Algorithms for Polarization Demultiplexing

Part of the content of this section has been published in [C6]:
J. Li, C. Schmidt-Langhorst, R. Schmogrow, D. Hillerkuss, M. Lauermann, M. Winter, K. Worms, C. Schubert, C. Koos, W. Freude, and J. Leuthold, "Self-coherent receiver for PolMUX coherent signals," in Proc. of Optical Fiber Communication Conference (OFC'11), Los Angeles (CA), USA, paper OWV5, 2011.

Different to the previous scheme, where external optical polarization controllers were needed in order to process PolMUX signals, in this section we will present a self-coherent receiver equipped with DSP algorithm which can demultiplex the two polarizations digitally. We experimentally validate the concept with 12.5 GBd (50 Gb/s) and 25 GBd (100 Gb/s) PolMUX-NRZ-QPSK signals.

4.2.1 DSP Outline

The receiver setup is similar as in Fig. 4.1(a). However, at the input of the receiver, the polarization states of the signals $E_{x,y}$ are unknown. They are projected onto the polarization axes of the receiver resulting in $E'_{x,y}$. The signals $E'_{x,y}$ are then sent into delay interferometers. At the output of the receiver, four electrical signals are received, which are the inphase and quadrature phase components of the signals combined with their delayed copy. The electrical signals are processed with DSP circuits.

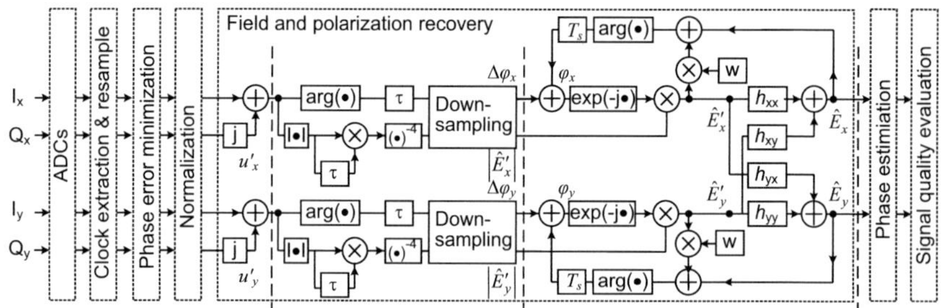

Fig. 4.4. Digital signal processing flow chart. Details of a field and polarization recovery algorithms are presented. ADC: analog-to-digital converter.

The DSP flow chart is shown in Fig. 4.4. After digitization of the electrical signals I_{Ix}, I_{Qx}, I_{Iy}, I_{Qy} the clock is extracted and four waveforms are synchronized and resampled with a time delay τ between adjacent samples. Then, two sources of phase errors in the DIs are numerically mitigated. One is the IQ-phase imbalance within each detected polarization, the other is the differential phase offset between the two detected polarizations. Both are static errors that can be measured in advance or estimated with the measured waveforms. After normalization the differential phasor $u'_{x,y} = I_{Ix,y} + jI_{Qx,y}$ is constructed to start field recovery. Its amplitude is estimated by $\left|\hat{E}'_{x,y}(t)\right| \approx \sqrt[4]{\left|u'_{x,y}(t)\right|\left|u'_{x,y}(t+\tau)\right|}$ [33]. Here the hat symbol $\hat{x}$ denotes an estimate of x. In the previous section we have shown that a partial time delay, e.g., $\tau = \frac{1}{2}\,T_s$ or $\frac{1}{4}\,T_s$ (T_s is the symbol duration) improves the estimation of the signal amplitude. However, the smallest τ is limited by the bandwidth of the photo detectors and analog-to-digital converters (ADCs). The estimated amplitude and the differential phase are then re-sampled to 1 sample per symbol. Next, the differential phase $\Delta\varphi_{x,\,y}$ is added to a phase reference and combined with $\left|\hat{E}'_{x,y}\right|$ resulting in the complex field phasor $\hat{E}'_{x,y}$. A constant modulus algorithm (CMA) equalizer [30] is subsequently applied for polarization demultiplexing. Details of the CMA equalizer are given in Appendix A.1.

However, the reconstructed phase of $\hat{E}'_{x,y}$ can be distorted by the accumulation of noise / error during the iterative process. The noise / error source includes thermal noise, quantization noise, and residual DI phase error etc. To illustrate this issue, we assume a simple DI phase error that deviate the detected differential phase from the original one by a constant offset $\varphi^{\varepsilon}_{x,y}$, so that $\Delta\varphi'_{x,y} = \Delta\varphi_{x,y} + \varphi^{\varepsilon}_{x,y}$. Putting this into Eq. (4.1.2), the phase of $\hat{E}'_{x,y}$ is,

$$\varphi_{x,y}(t_0 + n\tau) = \sum_{m=1}^{n}\left[\Delta\varphi_{x,y}\left(t_0 + m\tau\right)\right] + \varphi_{x,y}(t_0) + n\varphi^{\varepsilon}_{x,y} \qquad (4.2.1)$$

As a result the reconstructed phase will walk off with the increasing of time. In practice this walk off is different between the two polarizations as the phase noise / error at two polarizations are independent, see Fig. 4.5 (middle). From the right figure in Fig. 4.5, one can see that the reconstructed phases on the two polarizations deviate from their original phases with time increasing and especially the phase error difference between the two polarizations increases with time. This increasing phase error difference causes failure in the polarization demultiplexing algorithms.

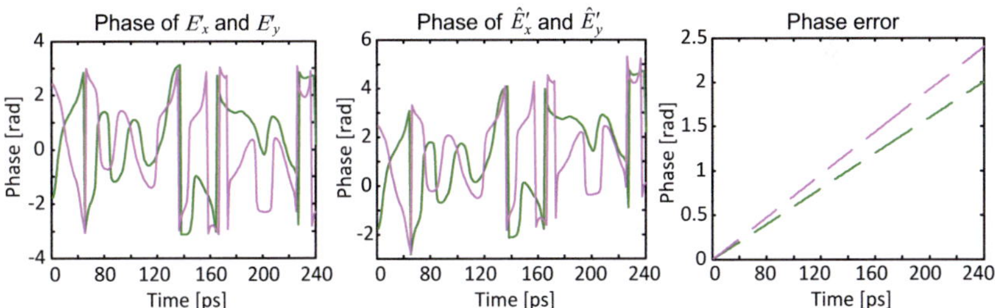

Fig. 4.5. Phase of the signals on two received polarizations. The left figure presents the phase of the original field E'_x (green) and E'_y (magenta) versus time. The figure in the middle presents the phase of the reconstructed field $\hat{E}'_{x,y}$ using Eq. (4.1.2). The right figure presents the phase error which is the difference between the phase of the original fields and the reconstructed fields.

Because the equalizer is always attempting to bring the signal towards the optimum state, we use the filtered field $\hat{E}_{x,y}$ as a corrector for the phase reference by combining it with the recovered phase. The smaller the phase error / noise is, the larger the weighting factor 'w' we can use. In Fig. 4.6, on the left side, it shows the phase error difference between the two polarizations varies with time significantly when there is no phase corrector turned on. When the phase corrector is switched on, the phase error difference increases very fast in the beginning, however, slows down and oscillates around a constant level after 100 symbols, as shown in Fig. 4.6 right. After the polarization and field recovery, $\hat{E}_{x,y}$ are sent into modules of phase estimation [78] and signal evaluation.

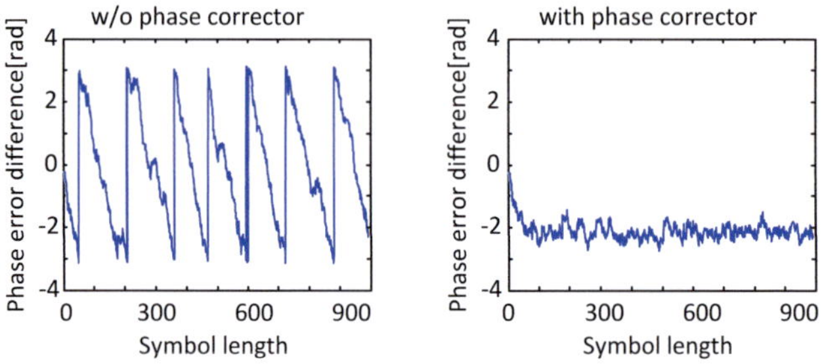

Fig. 4.6. Phase error difference between the two polarizations versus symbol length without phase corrector (left) and with phase corrector (right).

4.2.2 Experimental Setup and Results

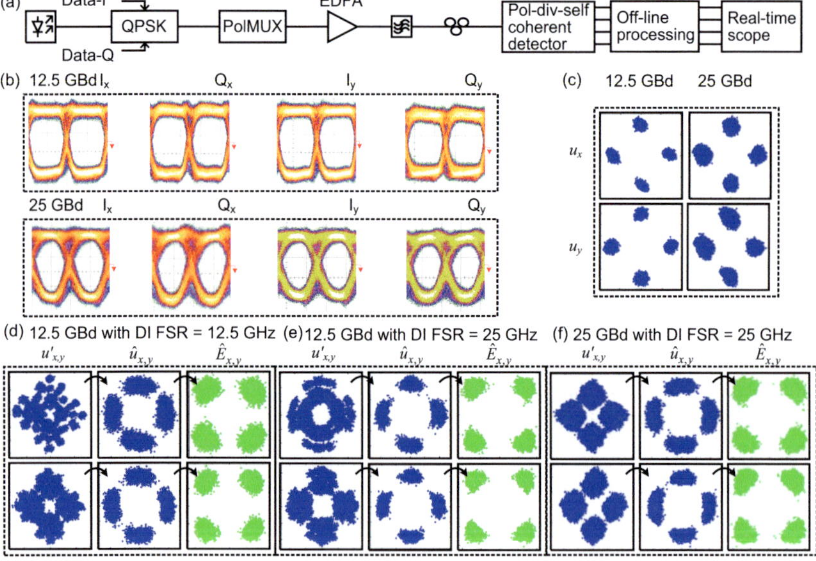

Fig. 4.7. (a) Experimental setup (b)+(c) Polarization-resolved I/Q eye diagrams and received constellations with receiver polarization aligned to the signal polarization; (d) - (f) Polarization-unaligned signal constellations (blue, differential phasors), constellations after polarization recovery (blue, differential phasors) and constellations after phase estimation (green, recovered field phasors), from left to right at (d) 12.5 GBd with DI FSR = 12.5 GHz, (e) FSR = 25 GHz and (f) 25 GBd with FSR = 25 GHz.

In the experiment (Fig. 4.7(a)) a PolMUX-NRZ-QPSK signal is generated by modulation of an external cavity laser (1547.6 nm) with two uncorrelated PRBS of length 2^{15}-1 in a dual-MZI modulator and a bit-aligned polarization multiplex stage with a delay of several bits. The signal is amplified and filtered before sent into the self-coherent receiver The input signal polarization needs only approximate adjustment to the receiver polarizations states (here, e.g., by a polarization controller) due to the CMA algorithm. A real time scope (50 GS/s, 20 GHz bandwidth) is used to digitalize the waveforms for off-line processing.

We used two signal baud rates, namely 12.5 GBd (50 Gb/s) and 25 GBd (100 Gb/s). As a reference we aligned the receiver polarization to the data signal, and time delay τ to the symbol duration T_s. The obtained eye diagrams (Fig. 4.7(b), measured with Agilent DCA sampling oscilloscope) have Q^2 factor ≈ 20 dB which verifies a good quality signal generation and self-coherent reception. Constellation diagrams of $u_{x,y}$ (Fig. 4.7(c), measured with real time scope) also verify a good quality reception. We attribute the stronger noise in the 25 GBd constellations to the bandwidth limitation of the real time scope.

Next we tested the self-coherent receiver with non-aligned polarizations. Since only polarization mixing is considered in this measurement, we set the FIR butterfly filter (CMA) length to 1 tap only although larger filter lengths could also be used to receive, e.g., signals with PMD. Exemplary results are shown in Fig. 4.7(d)-(f), where we plot the received constellation $u'_{x,y}(n)$, the constellation of $\hat{u}_{x,y}(n) = \hat{E}_{x,y}^{*}(n-1)\hat{E}_{x,y}(n)$ after polarization recovery and the recovered field constellation $\hat{E}_{x,y}(n)$ after phase estimation (window size of $3T_s$) for a 1 µs time duration. For the 12.5 GBd signal, a DI FSR of 25 GHz ($\tau = 40$ ps $= T_s/2$, Fig. 4.7(e)) yields a clearer recovered constellation (EVM = 13.0%, 12.4%) compared to a DI FSR of 12.5 GHz ($\tau = 80$ ps $= T_s$, Fig. 4.7(d), EVM = 19.0%, 17.5%) as predicted in the previous section. For reception of the 25 GBd signal (Fig. 4.7(f)) we only used a DI FSR of 25 GHz ($\tau = T_s$) due to the bandwidth limitation of the real time scope. In our experiments we found that the algorithm for polarization demultiplexing and field recovery converges well when the signal polarization is rotated against the PBS axes at the receiver input by < 35°. Simulations show that for angle rotations >35° (up to 45°), the recovered signal's EVM increases above 30%. We attribute this to two reasons. One is the rapidly varying amplitude which occurs in such cases on the detectors and which cannot be estimated with high enough accuracy by estimating the signal magnitude with a geometric mean as in Eq. (4.1.2). The other reason is the zero points when there is destructive interference between the signals on two polarizations.

4.2.3 Conclusion

We present a self-coherent receiver algorithm which relaxes the requirement of the polarization controller at the input of the receiver. By tuning the delay line in the DIs, we find a smaller delay would improve the quality of the reception. However, the algorithm cannot function at input signal with a fully mixed polarization. In the next section, we will further improve our algorithms and develop a self-coherent system that reconstruct the signal field with a better accuracy and is suitable for arbitrary polarization input.

4.3   A Self–coherent Receiver for Detection of PolMUX Coherent Signals

The content of this section is a direct copy of the Journal publication [J2]:
J. Li, R. Schmogrow, D. Hillerkuss, P. Schindler, M. Nazarathy, C. Schmidt-Langhorst, S.-B.
Ezra, C. Koos, W. Freude, and J. Leuthold, "A self-coherent receiver for PolMUX-signals,"
Opt. Express, vol. 20, no. 19, pp. 21413-21433, 2012.
Minor changes have been done to adjust the notations of variables and figure positions.

A self-coherent receiver capable of demultiplexing PolMUX-signals without an external polarization controller is presented. Training sequences are introduced to estimate the polarization rotation, and a decision feedback recursive algorithm mitigates the random walk of the recovered field. The concept is tested for a PolMUX-DQPSK modulation format where one polarization carries a normal DQPSK signal while the other polarization is encoded as a progressive phase-shift DQPSK signal. An experimental demonstration of the scheme for a 112 Gbit/s PolMUX-DQPSK signal is presented.

4.3.1 Introduction

Coherent receivers are key to next generation multi-level polarization-multiplexed modulation formats such as differential quadrature phase-shift keying (PolMUX-DQPSK) [26]. However, coherent detection requires a costly high-quality local oscillator (LO) laser. While coherent detection provides highest sensitivity the price of a narrow linewidth laser is not within reach for many applications such as in metro and access networks. A substitute that does not require a LO would thus be of interest.

Recently, self-coherent detection (SCD) using delay interferometers (DI) has been suggested as a substitute for coherent reception [32][33][76][77][J3][C6]. The concept extends self-homodyne differential detection of signals such as the differential phase-shift keying (DPSK) modulation format, where phase differences of zero or 180° are conveyed. SCD schemes do not need an additional laser as a LO, instead the delayed copy of the signal is used for detection. Compared to a coherent receiver that requires digital signal processing (DSP) for carrier phase and frequency estimation, SCD requires advanced DSPs to reconstruct both the phase and the amplitude of the optical field by comparing an actually received complex sample with a previously reconstructed sample. While self-coherent receivers are advantageous with respect to hardware complexity, they mainly suffer from two issues. Firstly, SCD usually shows an optical signal-to-noise ratio (OSNR) penalty compared to coherent receiver. Secondly, SCD has not yet been able to receive a PolMUX signal.

The optical signal-to-noise ratio (OSNR) penalty of SCD compared to coherent receiver is about 2.3 dB [24], due to the fact that the reconstruction process refers to previously estimated symbols which are also noisy. Especially the quantization noise of analog-to-digital converters (ADC) at the receiver is a degrading factor [J3]. This OSNR penalty may be reduced with proper DSP algorithms [33][54][J3]. These algorithms, which were developed for single polarization, are not directly applicable to self-coherent receivers for PolMUX signals, since the unknown crosstalk between the two polarizations at the receiver strongly distorts the signal constellations. In addition, the fields in the two orthogonal polarizations may destructively interfere at the receiver polarizer yielding zero samples for the reference, causing outages in the reconstruction algorithm operation.

Three principal methods of polarization demultiplexing (PolDEMUX) were mostly developed for differential detection, which can in principle be also applied to SCD. The first method resorts to an external polarization controller for manually [5][7] or automatically aligning the signal to the receiver polarization axis [8]–[10]. No special signal format is needed. While automatic alignment does not require fast DSP circuits, it relies on high-speed endless polarization controllers. The second method, suitable for return-to-zero (RZ) signals, consists of interleaving the RZ signals propagating in two orthogonal polarizations by introducing a time offset of half symbol duration. At the receiver, the signals are detected using polarization-insensitive decision circuitry operating at twice the symbol rate [11], requiring high-speed photo-detectors and electronics. The third method uses multiple DIs with a variant of the least mean square (LMS) algorithm [12]. In this scheme the authors not only differentially detect the signals in two orthogonal polarizations, but they also cross-couple the two polarizations optically so that the cross product between the two orthogonally polarized signals can be formed by DSP. The advantage is that PolDEMUX occurs at the symbol rate. The disadvantage is the more complex optical circuitry, and the cost for twice as many photodiodes and ADCs, relative to the first SCD method and to a coherent receiver.

Unfortunately, in SCD, post-detection noise accumulates in the field reconstruction recursion, leading to a random-walk process of the complex recovered signal [C6][J3] if no preventive measures are taken. Due to the independent noise random walks in the two polarizations, the PolDEMUX algorithms [30] commonly used in coherent detection system are not directly applicable for SCD system.

In this section, we present a SCD receiver based on the principle that is different from the above methods. It does not require polarization controller, time interleaving, and a doubled set of detectors. We estimate the change of the state of polarization with training sequences, then use decision feedback to reconstruct the field and to mitigate the noise-induced random walk that affects the usual PolDEMUX methods. To address "zero-reference" outages in one polarization, we use the signal in the other polarization to estimate the "lost" reference. To this end, a variant of the PolMUX-DQPSK format is introduced, whereby the constellations in both polarizations are rotated by 45° relative to each other. An experimental demonstration of the SCD dual polarization scheme is presented for 112 Gbit/s PolMUX-DQPSK reception.

4.3.2 PolMUX Transmission and Self–coherent Reception

A polarization multiplexed transmission system with a self-coherent receiver is presented in Fig. 4.8(a). Two equivalent configurations of the self-coherent receiver optical frontend (OFE) are considered: An OFE with two phase-offset delay interferometers (DI) as in Fig. 4.8(b), and an OFE with an optical hybrid, Fig. 4.8(c).

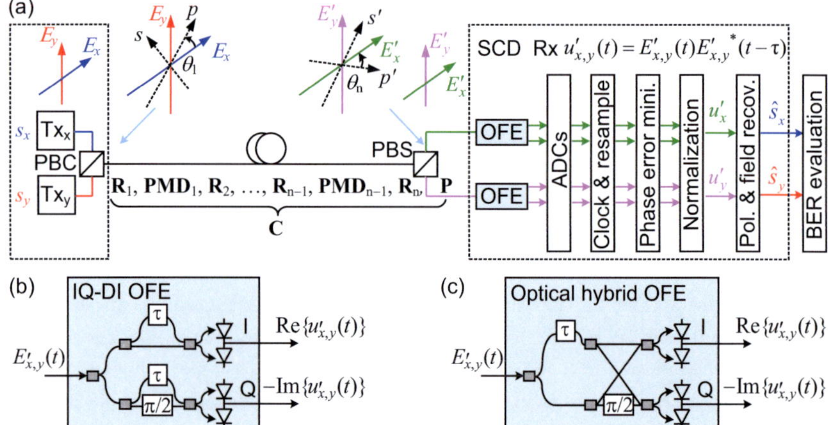

Fig. 4.8. Polarization multiplexed transmission system with self-coherent receiver (SCD Rx). (a) System schematic with SCD Rx and its DSP modules. Transmitters Tx_x and Tx_y generate two signals E_x and E_y carrying data s_x and s_y, which are then combined in a PBC to form a PolMUX signal. In the fiber the signal experiences a random change of the state of polarization described by a matrix $\mathbf{C}$ which consists of arbitrary polarization rotations $\mathbf{R_n}$, arbitrary order of $\mathbf{PMD_n}$, and arbitrary phase offset $\mathbf{P}$. The signal is then detected in a SCD Rx. There are two options for realizing the optical frontend (OFE) for detecting the inphase (I) and quadrature phase (Q) signal: (b) OFE with two quadrature phase-offset DIs (IQ-DI), (c) OFE with optical 90° hybrid.

The transmission link shown in Fig. 4.8(a) comprises two sub-transmitters (Tx_x, Tx_y) that carry two independent data streams, s_x and s_y. The signals generated by the two transmitters are orthogonally polarized. Their electric fields are denoted with E_x and E_y. The two signals are superimposed in a polarization beam combiner (PBC) to form the polarization multiplexed (PolMUX) signal, which is transmitted along a fiber, experiencing a random change of the state of polarization (SOP). The channel is modeled as a matrix $\mathbf{C}$ including arbitrary polarization rotations $\mathbf{R_n}$, arbitrary order of polarization mode dispersion ($\mathbf{PMD_n}$), and arbitrary phase offset $\mathbf{P}$, as detailed in Section 4.3.2.1. In the polarization diverse self-coherent receiver, the signal is split by a polarization beam splitter (PBS) in two orthogonal linear polarizations, the output fields of which are denoted with E'_x and E'_y. These signals are a mixture of the transmitted E_x and E_y as generated by the random SOP change along the fiber. Both polarizations are processed in a self-coherent receiver, the optical frontend (OFE) of which can be implemented by either of two following schemes.

Delay interferometer frontend. In Fig. 4.8(b) two DIs with balanced detectors are shown. The input of each DI is either $E'_x(t)$ or $E'_y(t)$, denoted by $E'_{x,y}(t)$ for short. Within each DI, the complex signal $E'_{x,y}(t)$ is mixed with its delayed copy $E'_{x,y}(t-\tau)$. By carefully choosing the excess phase difference between the two arms, either an inphase component $\mathrm{Re}\{u'_{x,y}(t)\} = \mathrm{Re}\{E'_{x,y}(t)E'_{x,y}{}^*(t-\tau)\}$ (I, zero phase difference, upper arm in Fig. 4.8(b)) or a quadrature phase signal $-\mathrm{Im}\{u'_{x,y}(t)\} = -\mathrm{Im}\{E'_{x,y}(t)E'_{x,y}{}^{*}(t-\tau)\}$ (Q, $\pi/2$ phase difference, lower arm in Fig. 4.8(b)) is detected by the balanced photodiode pair. We write the operation performed by the DI together with the balanced detector mathematically as

$$u'_{x,y}(t) = E'_{x,y}(t)E'_{x,y}{}^{*}(t-\tau). \tag{4.3.1}$$

The derivation of the equation can be found in Appendix A.3.

Optical hybrid frontend. An equivalent alternative scheme is shown in Fig. 4.8(c) where a 2×4 optical hybrid is used similarly to a coherent receiver. In lieu of the local oscillator, the incoming signal is delayed by τ, and this copy is fed into the 2×4 optical hybrid in order to interfere with the original signal. Multiple variants of 2×4 optical hybrids exist, implemented by discrete couplers, by star-couplers [36] or by multimode interference couplers [37]. The outputs are identical to the ones of the DI frontend.

After the balanced detection in either of the frontends, the complex quantities u'_x and u'_y are converted into the digital domain using analog to digital converters (ADCs). Digital signal processing (DSP) algorithms are then applied to recover the transmitted signals. At the outputs of the DSP, the demodulated signals $\hat{s}_x$ and $\hat{s}_y$ are available for further evaluation, for instance for a BER measurement. The DSP algorithms include clock recovery, resampling, phase error correction, normalization, polarization and field recovery. These last two functions are the subject of a new algorithm, which is our focus in this section.

In the remainder of this section we will present a model of the fiber channel, with which we analyze the interference between the Tx_x and Tx_y signals and summarize the challenges that need to solve. At the end, the principle of our transmitter and receiver will be presented.

4.3.2.1 Channel Model

A polarization multiplexed transmitter with a fiber channel and a polarization diverse SCD receiver is shown in Fig. 4.8(a). The fiber channel introduces some arbitrary SOP changes so that the detected polarization consists of a mixture between the two transmitted signals.

A possible scenario as of how the SOP may vary when passing from transmitter to receiver is depicted in the upper part of the figure. The quantities E_x and E_y represent the linearly polarized electric field components at the transmitter Tx_x and Tx_y. The two signals are combined at a PBC whose linear eigenstates are aligned with the polarization axes of E_x and E_y (blue and red coordinate systems). After the PBC, a fiber is attached. The fiber is modeled by numerous birefringent slices all of which have a different orientation of the fast and the slow axes and a different PMD. The first slice of the fiber model has linear eigenstates p (parallel) and s (senkrecht, perpendicular), which are rotated by an angle of θ_1 with respect to E_x and E_y. The signals have to be remapped to a new coordinate system. This operation can be described by a Jones matrix $\mathbf{R}_1$. The last slice of the fiber model has linear eigenstates p' and s'. At the end of the fiber a PBS with linear polarization eigenstates E'_x and E'_y projects this electric field on its eigenstates (green and magenta coordinate system), and in addition introduces a phase offset $\delta\phi_{\mathrm{PBS}}$ between the two signals. This phase offset is modeled by a matrix **P** as shown in Eq. (4.3.2). The rotation angle between the eigenstates of PBS and last fiber slice is θ_n. The coordinate system transformation is described by matrix $\mathbf{R}_n$.

Neglecting the propagator and any loss common to both polarizations, the channel is then described by a unitary Jones matrix **C** [33], which includes arbitrary polarization rotations $\mathbf{R}_n$, and arbitrary order of $\mathbf{PMD}_n$, and phase offset **P**,

$$\begin{bmatrix} E'_x \\ E'_y \end{bmatrix} = \begin{bmatrix} C_{11} & C_{12} \\ C_{21} & C_{22} \end{bmatrix} \begin{bmatrix} E_x \\ E_y \end{bmatrix}, \quad \mathbf{C} = \mathbf{P} \cdot \mathbf{R}_n \cdot \mathbf{PMD}_n \cdot \mathbf{R}_{n-1} \ldots \mathbf{R}_2 \cdot \mathbf{PMD}_1 \cdot \mathbf{R}_1,$$

$$\mathbf{R}_n = \begin{bmatrix} \cos(\theta_n) & \sin(\theta_n) \\ -\sin(\theta_n) & \cos(\theta_n) \end{bmatrix}, \mathbf{PMD}_n(\Delta\omega) = \begin{bmatrix} 1 & 0 \\ 0 & e^{-j(\Delta\omega\tau_{\mathrm{DGDn}}+\delta\phi_{\mathrm{PMDn}})} \end{bmatrix}, \mathbf{P} = \begin{bmatrix} 1 & 0 \\ 0 & e^{-j\delta\phi_{\mathrm{PBS}}} \end{bmatrix}. \tag{4.3.2}$$

Here, $\mathbf{PMD}_n$ represents the first-order PMD within the fiber. Assuming a time dependency $\exp(j\omega t)$ with optical angular frequency ω, the differential phase shift between the two principal states of polarizations in a signal bandwidth $\Delta\omega / (2\pi)$ is $\Delta\omega\,\tau_{\mathrm{DGDn}} + \delta\phi_{\mathrm{PMDn}}$, and the differential group delay is τ_{DGDn}. Additional effects could have been also included.

4.3.2.2 Challenges for Signal Recovery

The original Tx_x and Tx_y signals are undergoing SOP change in the channel. As a consequence the signals from Tx_x and Tx_y are mixed in the receiver. The SCD receiver needs to recover $\hat{s}_x$ and $\hat{s}_y$ from u'_x and u'_y. This can be done in three steps. The first step is the field recovery that E'_x and E'_y are reconstructed from u'_x and u'_y. The second step is the polarization recovery to recover E_x and E_y from E'_x and E'_y. At the end, it's the demodulation that $\hat{s}_x$ and $\hat{s}_y$ are retrieved from E_x and E_y. There are several challenges that need to be solved for such a SCD receiver.

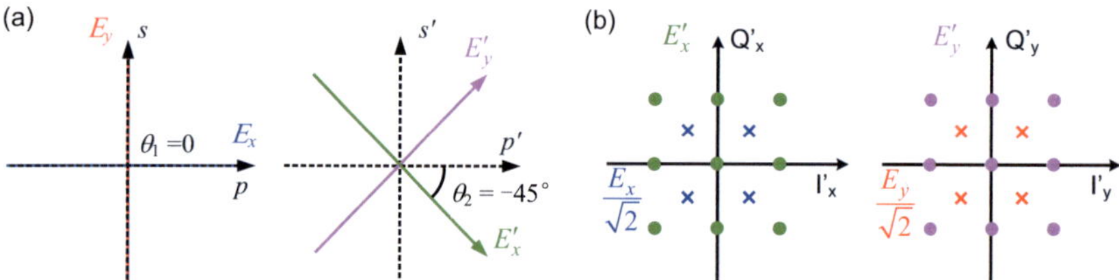

Fig. 4.9. (a) Illustration of Tx and Rx polarization states. (b) Constellation diagram as sent off by the Tx_x (×) and Tx_y (×) and constellation diagrams as received by the Rx in the two polarizations x' (•) and y' (•) when being sent over the channel in Eq. (4.3.3). The two QPSK (×, ×) are mixed and form new constellations $E'_x = (E_x - E_y)/\sqrt{2}$ (•) and $E'_y = (E_x + E_y)/\sqrt{2}$ (•). Zeros in the center of the constellations result from destructive interference.

The first challenge is that the mixing of the two signals could lead to destructive interference to the extent that the signal at the receiver becomes zero. To illustrate this issue we give an example using a PolMUX-QPSK signal transmitted in a very simple channel consisting only one fiber slice. In our simple channel illustration we neglect PMD and assume a zero phase offset after the PBS. In this case, the matrices $\mathbf{R}_1$, $\mathbf{PMD}_1$, and **P** are all identity matrices. The eigenstates of PBS with matrix $\mathbf{R}_2$ are rotated with respect to the eigenstates of

PBC, so that the transmitted field components E_x and E_y are combined to new electric field components E'_x and E'_y along the PBS eigenstates. Here we assume a rotation by $\theta_2 = -45°$, see Fig. 4.9(a). We find

$$\begin{bmatrix} E'_x \\ E'_y \end{bmatrix} = \frac{1}{\sqrt{2}} \begin{bmatrix} I'_x + jQ'_x \\ I'_y + jQ'_y \end{bmatrix} \exp(j\omega t) = \frac{1}{\sqrt{2}} \begin{bmatrix} 1 & -1 \\ 1 & 1 \end{bmatrix} \begin{bmatrix} E_x \\ E_y \end{bmatrix} = \frac{1}{\sqrt{2}} \begin{bmatrix} E_x - E_y \\ E_x + E_y \end{bmatrix}. \tag{4.3.3}$$

As an example, we assume QPSK where $E_{x,y} \in \frac{1}{\sqrt{2}}\{1+j, \; -1+j, \; 1-j, \; -1-j\}\exp(j\omega t)$. By substituting into Eq. (4.3.3), we arrive at the constellation diagrams in terms of the real (I, inphase) and imaginary part (Q, quadrature phase) of the complex envelope of the carrier $\exp(j\omega t)$, Fig. 4.9(b). As an issue with the reception scheme one may notice the zeros in the Rx constellation diagrams. The zeros result from a destructive interference between E_x and E_y when carrying identical symbols. These zeros result in outages of the field reconstruction for E'_x and E'_y. Special care need to be taken to overcome this issue (details in Section 4.3.2.5).

The second challenge is to avoid error propagation during the field reconstruction process. The error may propagate because the signal is reconstructed using the previous received symbol as a reference. With this a single error could lead to an error in all subsequently reconstructed fields. However, this would not be an issue if the value of the received quantities $s_{x,y}$ would not dependent on the previous detection but only on the difference between two adjacent samples of the field. Therefore, we differentially encoded the transmitter signals. The transmitter is going to be discussed in the next section and the error propagation will be discussed in more detail in Section 4.3.2.5.

The third challenge is to undo the SOP change of the fiber. For this we will introduce a training sequence to estimate the channel matrix **C** in the next section and Appendix A.4.

4.3.2.3 Transmitter with Differential Encoding and Training Sequence for Channel Estimation

The data in the transmitter are differentially precoded (DP) in order to avoid error propagation as outlined in Section 4.3.2.2. We start with the complex data sequence $s(n)$ belonging to a certain constellation. By differentially precoding, $s(n)$ is processed into a complex symbols $A(n)$ at discrete times $t_n = n\tau$ at multiples n of the symbol period τ [48],

$$A(n) = s(n)\breve{A}(n-1), \tag{4.3.4}$$

where $\breve{A}(n-1)$ retains the same phase as $A(n-1)$, but its magnitude is normalized to one, i. e., $\breve{A}(n-1) = A(n-1)/|A(n-1)|$. If we take standard DQPSK as an example, we have $s(n) \in \{\pm 1, \pm j\}$. Given the sequence $s(n) = [1, j, -1, 1, -j...]$ and $A(0) = 1$, we get $A(n) = [1, j, -j, -j, -1...]$.

The differentially precoded signal $A(n)$ is then modulated on an optical carrier. In a polarization multiplexed system we convey two signal streams $A_x(n)$ and $A_y(n)$ on the two orthogonal SOP, say, linear polarizations in x and y-direction. The optical signals after the modulation are denoted $E_x = A_x(n)\exp(j\omega t)$ and $E_y = A_y(n)\exp(j\omega t)$. The data flow is shown in Fig. 4.10. In practice, the oscillator is a laser and the mixer is implemented by an optical IQ-modulator.

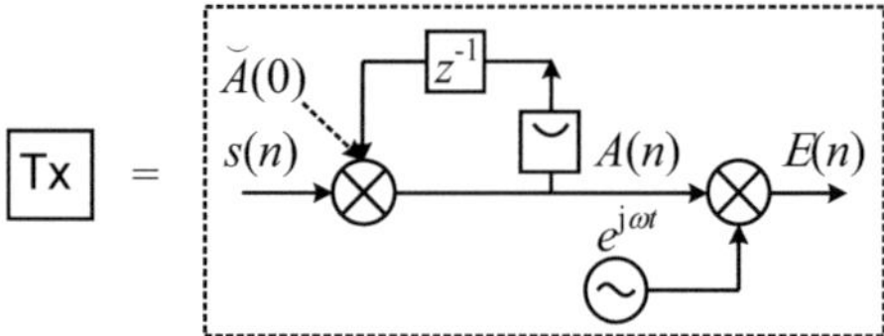

Fig. 4.10. Schematic drawing of the transmitter. $s(n)$ is differentially encoded into a $A(n)$ by adding $\breve{A}(n-1) = A(n-1)/|A(n-1)|$. The symbol z^{-1} (representing the z-transform) stands for a time delay by one bit. Then $A(n)$ is modulated on an optical carrier with angular frequency ω. The mixer output is the time sequence $E(n)$.

In the fiber channel, the signals experience an SOP change. Because in our case the optical fields $E'_{x,y}(t)$ cannot be measured directly, but rather $u'_{x,y}(t) = E'_{x,y}(t)E'_{x,y}{}^{*}(t-\tau)$ after the balanced detectors in Fig. 4.8(b), conventional polarization demultiplexing does not work, and we need a sequence of training symbols for estimating the channel matrix $\mathbf{C}$ which is determined by 4 complex quantities C_{ij} (8 real numbers). With known electric field components $E_x = A_x \exp(j\omega t)$ and $E_y = A_y \exp(j\omega t)$ and measured fields E'_x and E'_y, four values of $A_{x,y}$ in each polarization would suffice for uniquely calculating $\mathbf{C}$. However, we measure only the quantities $u'_{x,y}(t) = E'_{x,y}(t)E'_{x,y}{}^{*}(t-\tau)$, and therefore only phase differences can be determined. As a consequence, the estimated complex channel matrix coefficients $\hat{C}_{11}$ and $\hat{C}_{12}$ share a common undetermined phase factor so that only their phase difference is known. The same is true for $\hat{C}_{21}$ and $\hat{C}_{22}$. This does not affect the channel estimation (see Appendix A.4), but the number of determinable real quantities is reduced to 6. Therefore a minimum of three transitions per polarization has to be evaluated, leading to a minimum symbol number of 4. Longer symbol sequences could be employed to estimate $\mathbf{C}$ by minimizing the least-mean-square error for the matrix coefficients. For demonstrating the concept, we use the minimum training symbol length consisting of a preamble (0, 1, 1) and two groups of 4 complex symbols with three valid transitions for each polarization,

$$
\begin{array}{l}
A_x(1...8):...0 \quad 1 \quad 1 \,\big|\, 1 \quad 1 \quad 0 \quad 0 \,\big|\, A_x(8)... \\
\hline
A_y(1...8):...0 \quad 1 \quad 1 \,\big|\, 0 \quad 0 \quad 1 \quad 1 \,\big|\, A_y(8)...
\end{array}
\qquad (4.3.5)
$$

Any symbol from the transmitted constellation can be chosen for the symbols $A_x(8)$ and $A_y(8)$, which are used to start the field recovery algorithm. The preamble starting with the two zeros in the two polarizations serves as a uniquely identifiable symbol sequence which must not be part of the transmitted constellation. Appendix A.4 provides details of the channel estimation.

4.3.2.4 Self–coherent Receiver with Decision Feedback

The receiver recovers the transmitted information $s_{x,y}(n)$. The OFE of the self-coherent receiver processes the transmitted signal $E'_{x,y}(n)$ in polarization x' or y' by interfering them with their delayed copy as shown in Fig. 4.8. For simplicity, we replaced the SCD Rx, i.e., the mixed analog-digital circuit of the Rx (in Fig. 4.8) by an equivalent digital circuit (see Fig. 4.11). In Fig. 4.11, we take it for granted that a DSP unit transforms the analog inputs $E'_{x,y}(t)$ at sampling times $t = t_n = n\tau$ into digital quantities $E'_{x,y}(n)$.

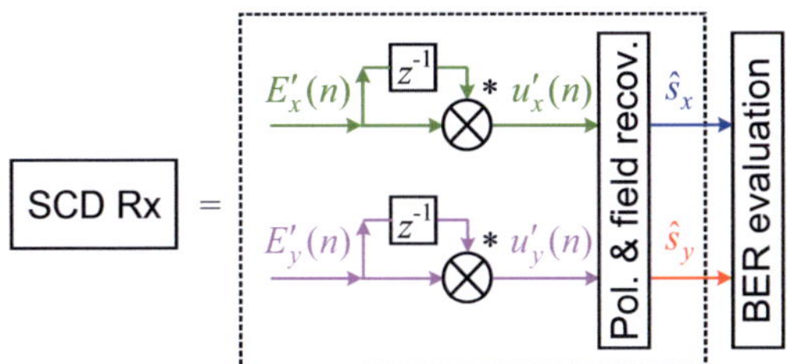

Fig. 4.11. Equivalent digital representation of the self-coherent detection receiver (SCD Rx). The received signal $E'_{x,y}(n)$ is combined with its conjugated delayed copy that generates $u'_{x,y}(n)$. The symbol z^{-1} (representing the z-transform) stands for a time delay by one bit.

If the SOP does not change, it is $E'_{x,y}(n) = E_{x,y}(n)$ and $u'_{x,y}(n) = u_{x,y}(n)$. After differential decoding we find

$$
\begin{aligned}
u'_{x,y}(n) = u_{x,y}(n) &= E_{x,y}{}^*(n-1)E_{x,y}(n) = A_{x,y}{}^*(n-1)A_{x,y}(n) \\
&= A_{x,y}{}^*(n-1)s_{x,y}(n)\breve{A}_{x,y}(n-1) = \left| A_{x,y}(n-1) \right| s_{x,y}(n),
\end{aligned}
\tag{4.3.6}
$$

where x^* denotes the complex conjugate of quantity x. It should be noted that $|A_{x,y}(n-1)| = 1$ can be chosen for signals $s_{x,y}(n)$ with constant modulus such as for DQPSK. Looking at Eq. (4.3.6) we see that with this normalization $u_{x,y}(n) = s_{x,y}(n)$, which means an ideal DQPSK detection. No polarization and field recovery algorithm is needed.

When the signal is transmitted through a non-ideal channel, polarization crosstalk may occur. Instead of $u'_{x,y}(n) = u_{x,y}(n)$ we receive

$$
\begin{aligned}
u'_x(n) &= E'_x{}^*(n-1)E'_x(n), \\
u'_y(n) &= E'_y{}^*(n-1)E'_y(n),
\end{aligned}
\tag{4.3.7}
$$

where E'_x and E'_y are the received electric fields for polarizations x and y, as described by the channel model of Eq. (4.3.2). The overall system is represented by the equivalent digital scheme in Fig. 4.12.

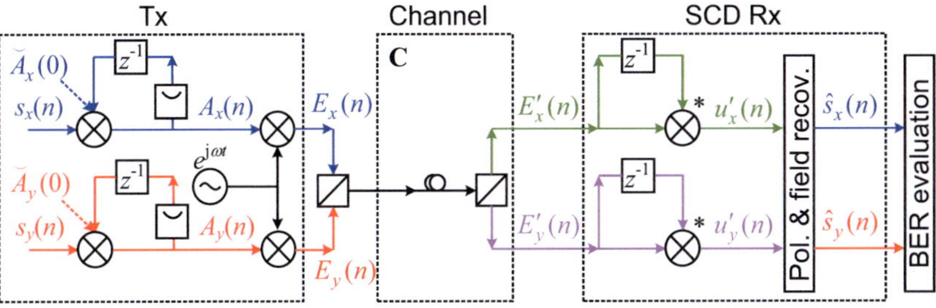

Fig. 4.12. Equivalent digital representation of the PolMUX signal transmission system. A PolMUX signal E_x and E_y carries encoded symbols of s_x and s_y. After the channel transmission, the receiver projects the signals on new polarization axes which results in E'_x and E'_y. In the SCD Rx, the signal interferes with its delayed copy generating quantities u'_x and u'_y. After polarization and field recovery the transmitted symbols s_x and s_y are recovered and sent for further evaluation.

At the receiver we measure u'_x and u'_y from which we want to recover the transmitted symbols s_x and s_y. However, these symbols are no longer simply obtained at the two DI outputs, due to polarization mixing. The challenge is to recover the fields $E'_x(n)$, $E'_y(n)$ and to perform polarization demultiplexing.

In a first step we need to derive E'_x and E'_y iteratively from the outputs u'_x and u'_y of the self-coherent receiver OFE. A recursive algorithm for single polarization field reconstruction alike to the one displayed in Fig. 4.13(a) was recently proposed in [J3]. Ideally, if we were to know an estimate $\hat{E}'_{x,y}(n-1)$ of the true symbol $E'_{x,y}(n-1)$, we easily could derive an estimate of the following symbol $E'_{x,y}(n)$ by solving $u'_{x,y}(n) = E'_{x,y}{}^{*}(n-1)E'_{x,y}(n)$ for the unknown $E'_{x,y}(n)$,

$$\hat{E}'_x(n) = u'_x(n)\big/\hat{E}'_x{}^{*}(n-1), \qquad \hat{E}'_y(n) = u'_y(n)\big/\hat{E}'_y{}^{*}(n-1). \tag{4.3.8}$$

The "hat" symbol $\hat{x}$ denotes an estimate of x. By repeatedly applying Eq. (4.3.8) we can recursively recover the field at sampling times t_n once we have an initial estimate $\hat{E}'_{x,y}(0)$.

In the second step a CMA (constant modulus algorithm) or decision directed LMS equalizer [30] can be applied for polarization demultiplexing same as in a coherent receiver. Once we know $\hat{E}_{x,y}(n)$, the symbols $\hat{s}_x$ and $\hat{s}_y$ are retrieved by differential detection, similarly to Eq. (4.3.6). The only change against Eq. (4.3.6) is that $\hat{E}_{x,y}(n-1)$ can now be normalized to one, because we are in the digital domain. Thus we directly obtain s_x and s_y,

$$\begin{aligned}
\hat{s}_x(n) &= \breve{\hat{E}}_x{}^{*}(n-1)\hat{E}_x(n) = \breve{A}_x{}^{*}(n-1)s_x(n)\breve{A}_x(n-1) = s_x(n), \\
\hat{s}_y(n) &= \breve{\hat{E}}_y{}^{*}(n-1)\hat{E}_y(n) = \breve{A}_y{}^{*}(n-1)s_y(n)\breve{A}_y(n-1) = s_y(n).
\end{aligned} \tag{4.3.9}$$

This equation indicates that $s_x(n)$ and $s_y(n)$ may be derived for any differentially encoding modulation format including QAM, and not only for constant-modulus signals.

However, the recursive algorithm in Eq. (4.3.8) has an issue with noise accumulation, because the signals are derived from the previous ones. The recovered field is perturbed by random walk-like noise. This leads to erroneous estimates even after a small number of samples [J3][C6]. Even without considering the analytic details, it is evident that as both $u'_{x,y}(n)$ and $\hat{E}'_{x,y}{}^{*}(n-1)$ are noisy, their ratio $\hat{E}'_{x,y}(n)$ is even noisier. Such independent noise random walks in the two polarizations make the PolDEMUX algorithms fail. To reduce the noise accumulation, training sequence is used in [J3] to monitor the random walk and reset the field recovery when the random walk goes beyond the limit. This method significantly reduces the random walk, however, cannot be simply applied for the polarization multiplexed signal because the training sequence can be strongly distorted due to polarization mixing. It also costs extra redundancy.

Another issue of the recursive algorithm in Eq. (4.3.8) is that because the initial estimate $\hat{E}'_{x,y}(0)$ is randomly chosen, the estimation can be wrong by a factor of $g = \hat{E}'_{x,y}(0)\big/E'_{x,y}(0)$. As a result, the odd sub-sequence would be scaled by $1/g^{*}$, $\hat{E}'_{x,y}(1,3,5...) = E'_{x,y}(1,3,5...)\big/g^{*}$, and the even sub-sequence would be scaled by g, $\hat{E}'_{x,y}(2,4,6...) = gE'_{x,y}(2,4,6...)$. The even and odd sub-sequences of the reconstructed signal would experience different scaling that need be rescaled afterwards [J3].

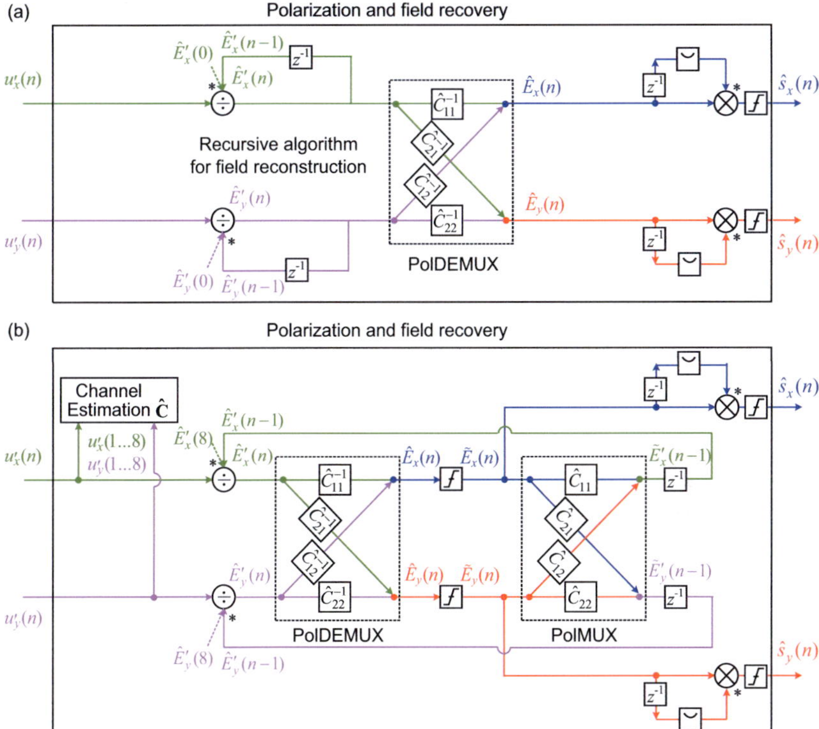

Fig. 4.13. Polarization and field recovery algorithm to derive the symbol s_x and s_y from complex signal $u'_{x,y}(n)$. (a) The conventional field recovery algorithm using solely Eq. (4.3.8). The recovered fields $\hat{E}'_{x,y}(n)$ are then sent to polarization demultiplexing (PolDEMUX) algorithm and differential decoder to retrieve s_x and s_y. This algorithm fails when there is noise accumulation. (b) New polarization and field recovery algorithms. It first performs the PolDEMUX, then uses decision circuits to remove noise. The 'clean' signal $\tilde{E}_{x,y}(n)$ is then multiplied with SOP change to generate the reference $\tilde{E}'_{x,y}(n-1)$ for the next field recovery.

In this section we advance [J3] by improving the reconstruction algorithm, mitigating the noise accumulation problem by means of a special decision feedback, see Fig. 4.13(b). The key idea is to avoid the use of the previous (noisy) symbol estimate $\hat{E}'_{x,y}(n-1)$ for computing the next one $\hat{E}'_{x,y}(n)$ as it was done in the Fig. 4.13(a). Instead, we perform the PolDEMUX first by multiplying the result with the inversion of the channel matrix **C**, and then make a decision $\tilde{E}_{x,y}(n)$ for the actual symbol thereby cleaning it from any noise. We then redo the PolMUX by multiplying $\tilde{E}_{x,y}(n)$ with the channel matrix **C** and introducing a delay by one bit. The resulting $\tilde{E}'_{x,y}(n-1)$ represents a noise-free reference for the next symbol, but can still be slightly inaccurate because of the imperfect channel estimation. However, an accumulation of noise is avoided. To further improve the accuracy of the channel estimation and to adapt to the channel variation over time it is possible to use the CMA or decision directed LMS equalizer [30] to adaptively adjust the channel matrix **C**. During the channel estimation, the initial symbols $A_{x,y}(8)$, which are translated into transmitted field quantities $E_{x,y}(8)$, lead us to the values of $\hat{E}'_{x,y}(8)$ as described in Appendix A.4. Thus $\hat{E}'_{x,y}(8)$ are used as the initial estimate for field reconstruction. At the end $\tilde{E}_{x,y}(n)$ are sent to differential decoder to retrieve $\hat{s}_{x,y}(n)$ same as in Eq. (4.3.9).

4.3.2.5 Field Outages caused by Polarization Crosstalk

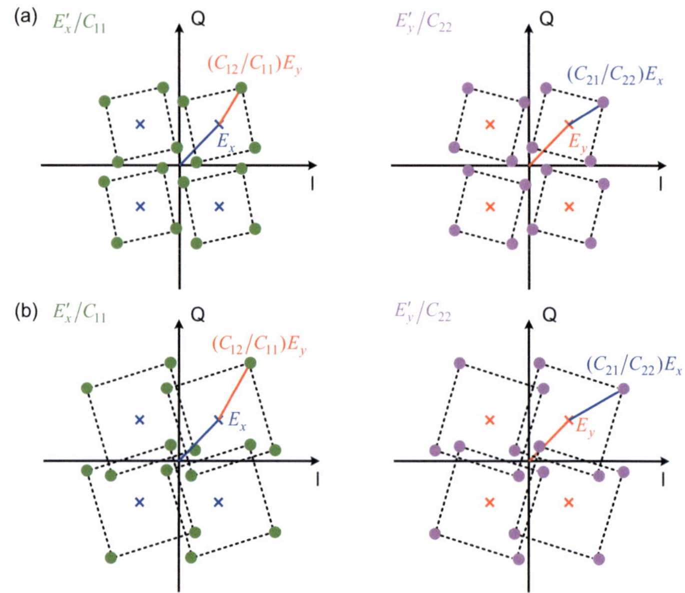

Fig. 4.14. Examples of PolMUX-DQPSK signal constellation diagrams wherein there is polarization mixing between the two signals. (a), the constellations of E'_x/C_{11} (•) or E'_y/C_{22} (•) have more weight from E_x (×) or E_y (×). (b), E'_x/C_{11} (•) or E'_y/C_{22} (•) have more weight from sub-constellation $(C_{12}/C_{11})E_y$ or $(C_{21}/C_{22})E_x$. Symbol combinations close to zero are detected that might cause outages in the field reconstruction.

In extreme cases, when the signals experience strong polarization crosstalk during transmission, one of the two signal polarization states at the receiver could fade due to destructive interference, so that $\hat{E}'_x(n-1)$ or $\hat{E}'_y(n-1)$ become zero. Applying Eq. (4.3.8) to recover $\hat{E}'_{x,y}(n)$ will then lead to an infinite output in one of the polarizations, and eventually generate an outage of the field reconstruction process. A non-zero very small value could either be quantized to zero or an inaccurate value could leads to wrong $\hat{E}'_{x,y}(n)$.

To mitigate this problem we modify the field recovery method. We consider a transmitted DQPSK constellation in both polarizations, each constellation may be represented by four phasors $E_{x,y}$ in the x and y-direction, respectively. After transmission, the fields E'_x and E'_y comprise elements with contributions from E_x and E_y. Using Eq. (4.3.2) we can express E'_x and E'_y as sum of the respective phasors, i.e., as $E'_x/C_{11} = E_x + (C_{12}/C_{11})E_y$ and $E'_y/C_{22} = E_y + (C_{21}/C_{22})E_x$. For DQPSK where each $E_{x,y}$ has four possible states we get a total of 16 constellation points for E'_x and E'_y, see Fig. 4.14. If the sub-constellations given by $(C_{12}/C_{11})E_y$ or $(C_{21}/C_{22})E_x$ in Fig. 4.14(a) and (b) are separated so that different sub-constellations points do not coincide, knowledge of $\mathbf{C}$ and one polarization state only (i. e., E'_x or E'_y) suffices to identify E'_y and E'_x due to the finite amount of possible states $E'_{x,y}$.

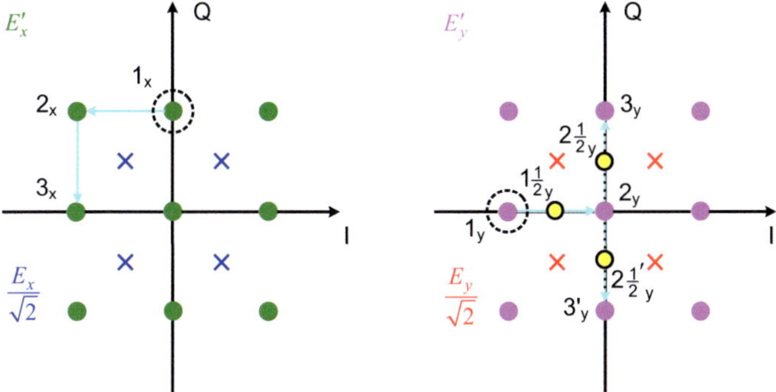

Fig. 4.15. Example of a PolMUX-DQPSK signal constellations when the linear transformation $\mathbf{C}$ in the fiber consists of an exact $-45°$ rotation. In the constellation of E'_x ($\bullet$), the signal transits from 1_x to 2_x to the ambiguous point 3_x. In the constellation of E'_y ($\bullet$), the signal transits from 1_y to 2_y to 3_y or $3'_y$ and the yellow points $1\frac{1}{2}{}_y$, $2\frac{1}{2}{}_y$, and $2\frac{1}{2}'{}_y$ are the halfway samples.

However, a second speciality needs to be taken into account. As soon as sub-constellation points coincide, they could result from more than one pair of E_x and E_y. In this case, if for example E'_y cannot be reconstructed due to an outage as in Fig. 4.9(b), the quantity E'_x cannot be used to uniquely determine E'_y. This problem can be solved by sampling not only the data points, but also halfways during the transition from one symbol to the next, i. e., by a twofold oversampling. For example, in the constellation diagram Fig. 4.15, the field E'_y transits from point $E'_y(1) \hateq 1_y$ (marked with dashed circle) to $E'_y(2) \hateq 2_y$ in the center of the constellation, so the next value $E'_y(3)$ could not be recovered because $E'_y(2)=0$ holds. Therefore we have to rely on the information E'_x only. However, $E'_x(3)=[E_x(3)-E_y(3)]/\sqrt{2}$ $=-1(e^{j\omega t})$ from Eq. (4.3.3) could result from either $E_x(3)=(-1+j)e^{j\omega t}/\sqrt{2}$ and $E_y(3)=(1+j)e^{j\omega t}/\sqrt{2}$, or, alternatively, from $E_x(3)=(-1-j)e^{j\omega t}/\sqrt{2}$ and $E_y(3)= (1-j)e^{j\omega t}/\sqrt{2}$. This ambiguity can be resolved by a twofold oversampling. If it is as assumed above $E'_y(1) \neq 0$ and $E'_y(2)=0$, then $E'_y(1\frac{1}{2}) \neq 0$ would hold, so $E'_y(2\frac{1}{2})$ can be calculated with Eq. (4.3.8). Depending on the result ($E'_y(2\frac{1}{2})=\pm j(e^{j\omega t})/2$), we decide on the nearest neighbor $E'_y(3)=[E_x(3)+E_y(3)]/\sqrt{2}=\pm j(e^{j\omega t})$, see Eq. (4.3.3). Now the quantities $E_{x,y}(3)$ can be uniquely determined.

A third special case needs consideration. If two values $E'_y(n)=0$ and $E'_y(n+1)=0$ follow each other, i. e., if no transition takes place, then also $E'_y(n+\frac{1}{2})=0$ is true. As a consequence, $E'_y(n+1\frac{1}{2})$ cannot be recovered so that the oversampling method fails. Therefore such sequences of zeros have to be avoided from the very beginning. To this end, we transmit one of the polarizations, say the y-polarization, with a DQPSK modulation and a progressive $45°$ phase shift being added for consecutive symbols E_y, Fig. 4.16 right column. This could be done with an additional clocked phase modulator (details in Appendix A.5). The resulting DQPSK phases for the symbols s_y are 45°, 135°, −135° and −45° [79]. The other polarization transmits a normal DQPSK signal with phases 0°, 90°, 180° and −90° for the symbols s_x, Fig. 4.16 left column. Looking at Fig. 4.16 and assuming that $E_x(1) = (1+\mathrm{j})e^{\mathrm{j}\omega t}/\sqrt{2}$ destructively interferes with $E_y(1) = (1+\mathrm{j})e^{\mathrm{j}\omega t}/\sqrt{2}$, none of the 4 possibly following values $E_y(2) = \pm 1(e^{\mathrm{j}\omega t}), \pm\mathrm{j}(e^{\mathrm{j}\omega t})$ could destructively interfere with any of the constellation points $E_x(2) = (\pm 1 \pm \mathrm{j})e^{\mathrm{j}\omega t}/\sqrt{2}$. Thus, series of consecutive zeros for $E'_{x,y}$ cannot occur.

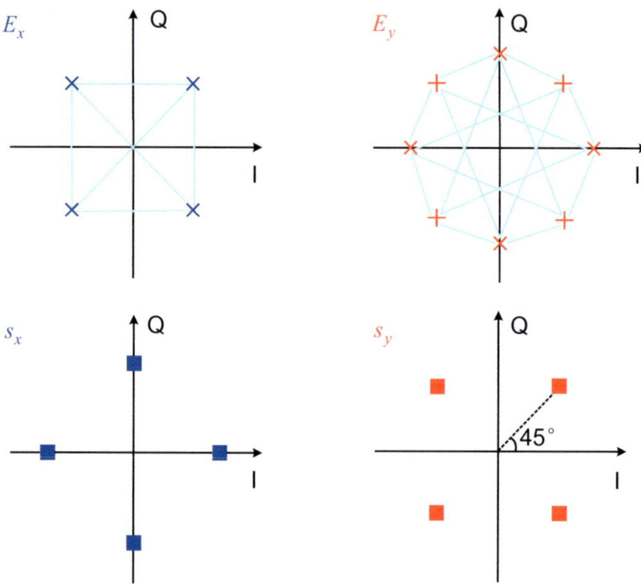

Fig. 4.16. PolMUX-DQPSK signal constellation diagrams. Top row shows the phasor of the signals (E_x and E_y) and their transition lines, the bottom row shows the phasors of the encoded symbols (s_x and s_y) which are also the transitions of the signals (E_x and E_y).

Finally, a fourth case needs consideration. If a wrong decision is taken by the decider circuit in Fig. 4.13(b), the outcome $\tilde{E}_{x,y}(n)$ is erroneous. For a constant modulus DQPSK signal, the wrongly detected field can be written as the correct field times a phase factor, $\tilde{E}_{x,y}(n) = E_{x,y}(n)\exp(\mathrm{j}\varphi^{\varepsilon}_{x,y})$, where $\exp(\mathrm{j}\varphi^{\varepsilon}_{x,y}) \in \{\pm 1, \pm \mathrm{j}\}$. This error carries over to the next reference for the field recovery step $\tilde{E}'_x(n-1) = \exp(\mathrm{j}\varphi^{\varepsilon}_x)\hat{C}_{11}E_x(n-1) + \exp(\mathrm{j}\varphi^{\varepsilon}_y)\hat{C}_{12}E_y(n-1)$ and $\tilde{E}'_y(n-1) = \exp(\mathrm{j}\varphi^{\varepsilon}_x)\hat{C}_{21}E_x(n-1) + \exp(\mathrm{j}\varphi^{\varepsilon}_y)\hat{C}_{22}E_y(n-1)$. If the polarization crosstalk is small, i. e., if $\hat{C}_{12}, \hat{C}_{21}$ are sufficiently small, the error would not propagate due to the differential phase detection scheme: In Eq. (4.3.8) the new field $\hat{E}'_{x,y}(n)$ has the same phase factor $\angle\left(1/\hat{E}'_{x,y}{}^{*}(n-1)\right) \propto \angle\hat{E}'_{x,y}(n-1)$ as for the previous field estimate, and differential phase detection then removes this error. However, if polarization crosstalk is strong, the generally different erroneous phase factors from both polarizations appear in mixed form and do not cancel out by differential detection. In the unlikely case that the error phases in both polarizations are the same, $\varphi_x^\varepsilon = \varphi_y^\varepsilon$, the reference fields in both polarizations would be rotated by the same amount, and the situation would be as in the case of weak polarization coupling. However, if $\varphi_x^\varepsilon \neq \varphi_y^\varepsilon$, the error propagates and leads to very inaccurate estimates $\hat{E}'_{x,y}(n)$. After polarization demultiplexing, large error vectors magnitudes $|e_{x,y}(n)| = |\tilde{E}_{x,y}(n) - \hat{E}_{x,y}(n)|$ would then be found after the decision circuit, and the subsequence field recovery would also become faulty. When this happens, we could minimize the error vectors magnitudes by replacing $\hat{E}'_{x,y}(n)$ with a proper choice out of the possible 16 constellation points at each polarization such that the erroneous phases in both polarizations become the same and $\varphi_x^\varepsilon = \varphi_y^\varepsilon$ holds, a situation which was discussed previously. The nth estimates may be still in error, however, after a maximum of two erroneously detected symbols the error propagation stops and the algorithm turns back to normal operation.

4.3.3 Experimental Setup and Results

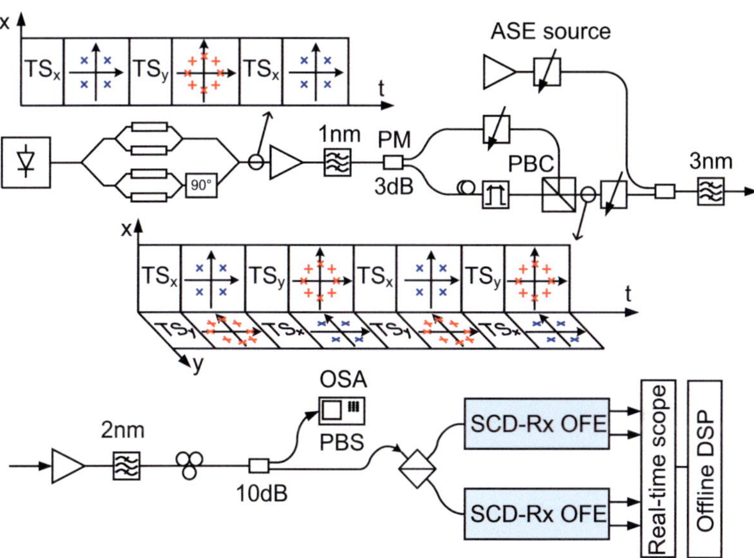

Fig. 4.17. Experimental setup, the signal is generated in a software defined Tx as a series of frames. In each frame, there is a training sequence (TS$_{x,y}$) for one polarization followed with a DQPSK signal or a progressive-phase DQPSK signal. Then the signal is split and combined in a PBC to form a PolMUX signal. The signal is then experienced with an arbitrary polarization rotation and sent to the self-coherent receiver. An ASE source is used to emulate different OSNR levels.

The experimental setup is depicted in Fig. 4.17. A 28 GBd NRZ-DQPSK signal is generated by modulating an external cavity laser (wavelength 1550 nm) with two uncorrelated data sequences applied to an IQ-modulator. The data sequences are generated by a software-defined transmitter [42]. To encode the two polarizations with the normal and the progressive-phase DQPSK signal using a single modulator, we use a special polarization multiplexing emulator. A first frame is generated with normal DQPSK symbols. It consists of the training sequence (TS$_x$) of the first polarization (Section 4.3.2.3) and a DQPSK signal with a PRBS sequence of length $2^8 - 1$. We limit the combination of the training sequence and the DQPSK sequence to a frame of 128 symbols. The next frame contains the symbols of the progressive-phase DQPSK signals, which has a constellation diagram with 8 phase states, however, only 4 possible phase states in one symbol period as discussed in Fig. 4.16. The frame comprises the training sequence (TS$_y$) of the other polarization and a PRBS sequence of length $2^8 - 1$ that is offset from the previous PRBS by 10 bits. The frame is limited to 128 symbols as before. Thus after the IQ-modulator, we have a data sequence of adjacent frames for the two polarizations. The DSP in the receiver has to be switched according to the rhythm of the modulation frame exchange.

The signal is then amplified with a polarization maintaining (PM) EDFA and filtered with a 1 nm filter. Then we use a 3 dB PM coupler to split the signal onto two arms, where one arm is delayed by 128 symbols with respect to the other one. Both arms are then combined. This yields a 112 Gbit/s bit-aligned PolMUX signal with training sequences for the two polarizations with alternating NRZ-DQPSK and progressive-phase NRZ-DQPSK signal frames. To mimic signals with different OSNR, an amplified spontaneous emission (ASE) noise source has been added. The signal is then filtered with a 3 nm filter and re-amplified before being sent to the self-coherent receiver. At the receiver input we use a 2 nm filter to further suppress the out-of-band noise. A polarization controller emulates polarization rotation in the fiber link. The receiver comprises two pairs of IQ delay interferometers as in Fig. 4.8(b). The DI delay times equal one symbol duration. After the balanced photo-detectors, the signal is captured by the real-time scope of an Agilent optical modulation analyzer (80 GS/s, 32 GHz bandwidth), which samples the waveforms for off-line processing.

The off-line processing consists of various steps. First, squaring clock recovery algorithm [80] is applied to down-sample the signal to two samples for each inphase and quadrature phase components of $u'_{x,y}(n)$. Then the signals are normalized and combined to form the complex representations $u'_{x,y}(n)$. After digitally removing the static phase errors in the delay interferometers, we apply our field and polarization recovery algorithm described in Section 4.3.2.4. The channel estimation is performed on the first 128 symbols, and afterwards is discontinued, unless it becomes necessary to re-estimate the channel.

We first test the algorithms with low ASE noise. In Fig. 4.18, we present signal reception under 6 different polarization rotations. Constellation diagrams of the signals $u'_{x,y}(n)$ before performing the polarization recovery algorithms are presented on the left side of each set. Because the signal is repeated with a normal NRZ-DQPSK and a 45° offset NRZ-DQPSK, the constellation resembles an 8PSK format. The zeros in the center are part of the training sequence. During field and polarization recovery, the training sequence is removed from the data and the 45° offset is also canceled. The estimated $\hat{E}_{x,y}$-constellations after field and polarization recovery are shown on the right side of each set. The recovered signals' error vector magnitude (EVM) are all below 16%, corresponding to BER less than 10^{-9} [81].

We then tested the algorithms for different OSNR values and polarization states with and without field and polarization recovery algorithm, Fig. 4.19. A maximal recording time of 350 μs (98×10^5 symbols) has been used for the BER evaluation. For polarization states 'D' and 'E' without recovery algorithm, the BER at OSNR 16 dB is anomalously high. This could be due to the failure of the clock recovery algorithm at low OSNR.

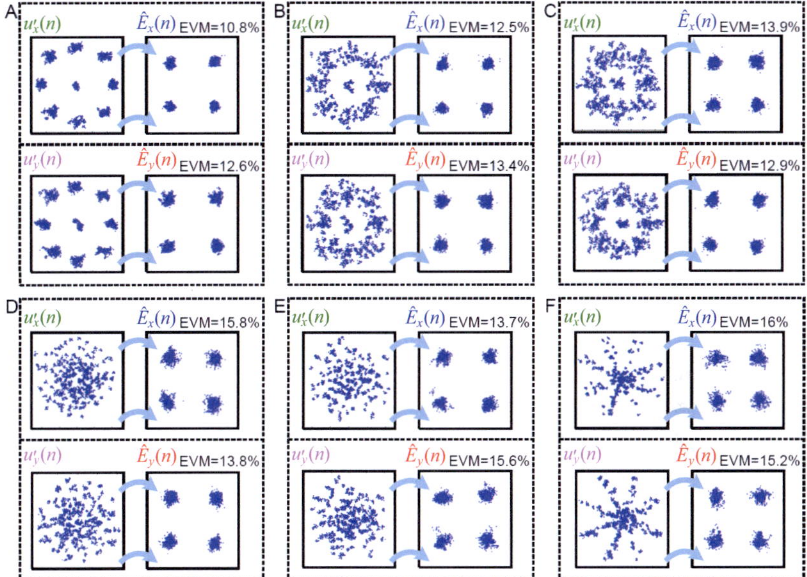

Fig. 4.18. Constellation diagrams of detected signals $u'_{x,y}$ in x and y polarizations together with the recovered symbols $\hat{E}_{x,y}$ for 6 different polarization states. All measurements were performed at low ASE noise, and 3072 symbols were evaluated.

For comparison, we detected the polarization-aligned signal with both a homodyne coherent receiver (black symbols) and with the self-coherent receiver (red symbols). For the signals detected with the homodyne receiver we applied a Viterbi-Viterbi algorithm [82] to correct the phase drift of the local oscillator.

To test the SCD receiver with increasing ASE the same 6 different polarization rotations have been tested as above. When the polarization is aligned, we get the best performance – with and without polarization recovery. The result is virtually same to the one from the coherent receiver. Theoretically, the coherent receiver should provide a result that is about 2.3 dB better than with the differential detection [24]. However, our coherent receiver is not ideal, and other researchers reported similar performance at almost the same symbol rate [26].

It can be seen that our novel algorithm substantially improves the self-coherent detection scheme. While the signal cannot be recovered for a 45° polarization rotation ((F) in Fig. 4.18 and Fig. 4.19) without our algorithm, the signal can be recovered with the recovery algorithm in any polarization state.

When the polarization crosstalk is strong we observe an OSNR penalty. This could be due to the fact that the recovered signal is calculated with the signal on two polarizations therefore noise from two sources contributes to the signal: The larger the polarization crosstalk, the higher the noise penalty. When there are (close) zeros at one polarization, the demodulation is then based on the other polarization which comprises 16 symbol constellation points. This also introduces a higher OSNR penalty to the signal. We should also consider the limitation of the equipment used in the experiment, i.e., the sampling rate and bandwidth of the real-time scope is not sufficient to fully track the transitions of the signal. Therefore the method described in Section 4.3.2.5 can only operate with limited efficiency.

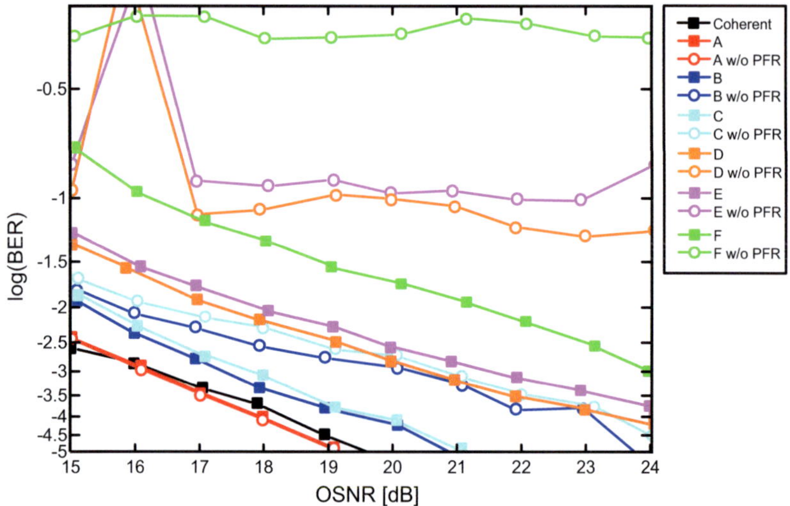

Fig. 4.19. BER versus OSNR plots of self-coherent receiver for 6 different polarization states with polarization and field recovery (PFR) algorithm (solid symbols) and without PFR algorithm (empty symbols). The black curve shows the OSNR versus BER for a coherent receiver.

4.3.4 Conclusion

We presented a self-coherent receiver operating without an external polarization controller. The self-coherent frontend consists of a pair IQ delay interferometers, or it uses a conventional optical hybrid with the LO replaced by a delayed version of the received light for each of the two polarizations. The DSP features a new algorithm based on training sequences to estimate the change of the state of polarization, and a decision feedback to mitigate the noise-driven random walk of the recovered field. The concept was tested for a PolMUX-DQPSK modulation wherein one polarization transmits a normal DQPSK signal, while the other one transmits a progressive-phase DQPSK signal. We experimentally demonstrated that this self-coherent receiver is capable of demultiplexing signals with arbitrary polarizations. In particular, we successfully tested the scheme on a 112 Gbit/s PolMUX-DQPSK signal under different polarization states at different OSNR levels.

5 Applications of Delay Interferometer

Delay interferometer has various applications in optical communications technology. Besides their application in the self-coherent [J2][J3][C6][C13][C19][C21] and differential direct detection [J5][J6][J8][C1][C2][C3][C8][C9][C10][C18] as discussed previously in this thesis, delay interferometers developed during this thesis have been used in reception of optical OFDM signal and all optical wavelength conversion.

The reception of optical OFDM signal is based on an optical fast Fourier transform (FFT) circuit. It is achieved by cascading several delay interferometers with careful chosen delay time [J4][J7][C7][C11][C12][C14][C16][C17].

The all optical wavelength conversion is combining a delay interferometer together with a semiconductor optical amplifier (SOA). The signal on one carrier is converted onto a second carrier based on the nonlinear effects of SOA. The delay interferometers together with band-pass filters reshape the pulses back to non-inverted [J9][C20][C22][C24][C25][74].

Additionally, cascading delay interferometers can offer various signal processing functions. One important application is for chromatic dispersion compensation [83][84].

In this chapter, a brief overview of the principle and experimental demonstration of optical FFT is firstly presented. Part of the content has been published in OFC 2010 conference proceeding [C16]. Then the principle and demonstration of all wavelength conversion is given. This part of the thesis has been published in Journal of Lightwave Technology [J9], CLEO conference proceeding [C25] and ICTON conference proceeding [C24]. At the end, the principle of chromatic dispersion compensation using DIs is briefly presented.

5.1 Optical Fast Fourier transforms (FFT)

Part of the content of this section has been published in [C16]:
D. Hillerkuss, A. Marculescu, J. Li, M. Teschke, G. Sigurdsson, K. Worms, S. Ben Ezra, N. Narkiss, W. Freude, and J. Leuthold, "Novel optical fast Fourier transform scheme enabling real-time OFDM processing at 392 Gbit/s and beyond," in Proc. of Optical Fiber Communication Conference (OFC'10), San Diego (CA), USA, paper OWW3, 2010 [Best Student Paper Award].

The principle of OFDM signal has been explained in Section 2.1.5. At the receiver a FFT circuit is necessary to separate the subcarriers. Conventionally the signal is first captured by a coherent receiver and then converted into digital domain. FFT is then performed electronically. While electronic signal processing is possible at 10 Gbit/s and potentially at 40 Gbit/s, future core channel bitrates are expected to operate at 100 Gbit/s and up to 1 Tbit/s [85], where electronic processing is difficult and severe power consumption issues arise. An optical implementation of the FFT has been demonstrated by Sanjoh et al. [86]. Here, an $N \times N$ waveguide grating router arrangement has been chosen. A number of N phase stabilization elements were simultaneously optimized with respect to their relative phases. While this is an interesting approach, it still requires the stabilization of N phases, and the complexity is unnecessarily high if, e.g., only one subcarrier is to be received.

A simple all-optical implementation of a FFT algorithm performing simultaneous serial-to-parallel (S/P) conversion and FFT calculation within a cascade of delay interferometers (DIs) has been introduced by our group recently [J4][J7][C7][C11][C12][C14][C16][C17]. The new scheme requires only passive DIs (as in all such schemes, time gating / sampling elements are required in addition), it scales well with the bitrate (enabling Tbit/s-OFDM), and it requires a very small amount of energy.

5.1.1 Operation Principle

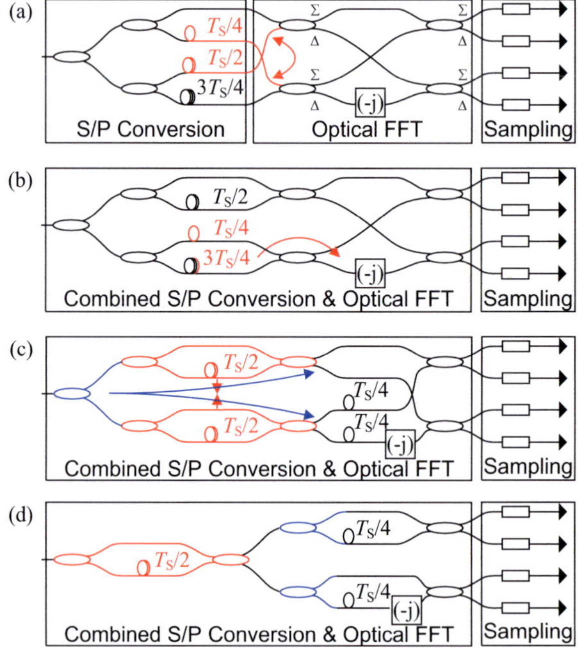

Fig. 5.1. Exemplary 4-point optical FFT for OFDM symbol duration T_s with S/P conversion, FFT and sampling. (a) Traditional implementation [88]; (b) Two paths are switched [89]; (c) An equivalent structure consisting of two delay interferometers with the same differential delay. The additional $T_s/4$ delay is moved out of the second DI; (d) Two identical DIs are replaced by a single DI followed by signal splitters. This scheme represents our new simplified S/P conversion and FFT scheme.

The fast Fourier transform (FFT) is an efficient method to calculate the discrete Fourier transform (DFT). For a number of time samples N, DFT is given as,

$$X_m = \sum_{n=0}^{N-1} \exp\left[-j2\pi \frac{mn}{N}\right] x_n, \quad m = 0, ..., N-1, \tag{5.1.1}$$

where $N=2^p$, with p being an integer. It transforms the N inputs x_n into N outputs X_m. If the x_n represent a time-series of equidistant signal samples of signal $x(t)$ over a time period T, the X_m will be the unique complex spectral components of signal x repeated with period T. The FFT typically "decimates" a DFT of size N into two interleaved DFTs of size $N/2$ in a number of recursive stages [87] so that

$$X_m = \begin{cases} X_{E_m} + \exp\left[-\mathrm{j}\dfrac{2\pi}{N}m\right]X_{O_m} & \text{if } m < \dfrac{N}{2}, \\[2em] X_{E_{m-N/2}} - \exp\left[-\mathrm{j}\dfrac{2\pi}{N}\left(m-\dfrac{N}{2}\right)\right]X_{O_{m-N/2}} & \text{if } m \geq \dfrac{N}{2}. \end{cases} \tag{5.1.2}$$

The quantities X_{E_m} and X_{O_m} are the even and odd DFT of size $N/2$ for even and odd inputs x_{2l} and x_{2l+1} ($l = 0, 1, 2,...N/2 - 1$), respectively.

Marhic [88] has shown a possible implementation of an optical circuit which performs an FFT. We depict this solution in Fig. 5.1(a) for $N = 4$. The S/P conversion, FFT processing and time gating provide the frequency-samples X_k. The number of couplers is the complexity, $C_{std} = N-1+(N/2)\log_2 N$. The number of phase shifters that need to be stabilized with respect to each other is $N \log_2(N)$. An implementation of the Marhic approach for $N = 4$ needs a total number of 7 optical couplers and 8 differentially phase-stabilized arms. The scaling number of the couplers and phase shifters renders the scheme impractical for large N. Yet, a simplification is possible. After combining and rearranging the elements of the S/P and FFT stage (see Fig. 5.1(b, c, d)), a functionally equivalent structure results. The new optical FFT processor consists only of $N - 1$ cascaded DIs with a small complexity of only $C_{DI} = 2(N - 1)$ couplers, where $C_{DI} \leq C_{std} \ \forall \ N$. Also, in this implementation only the phase of $N - 1$ DIs needs stabilization, and no inter-DI phase adjustment is required. The subsequent time gates define the FFT window. It has to be aligned to the OFDM symbols for suppressing inter-symbol and inter-carrier interference. For $N = 4$ the new solution requires only 3 phase stabilizations and 6 couplers.

An additional simplification is obtained for $N > 4$. It can be shown that the first two DI stages have the largest impact on the overall performance, so that only two DI stages are needed, and any additional DI can be replaced by passive splitters and band-pass filters thereby simplifying the FFT processing even further. It needs to be clarified, that for $N = 2$ our new implementation and the implementation by Marhic [88][89] lead to an identical structure. Our novel FFT scheme has a reduced complexity $C_{DI} < C_{std}$ for $N > 2$, or if only a subset of subcarriers is to be received.

In the following, we demonstrate an FFT for $N = 9$ subcarriers using only the first two stages of the DI cascade together with a band-pass filter for selecting the subcarrier of interest.

5.1.2 Experimental Implementation and Results

The OFDM receiver-transmitter pair is shown in Fig. 5.2. The transmitter is quite similar to the coherent-WDM transmitter proposed in [90] but it does not require any phase stabilization. It employs a 50 GHz optical comb generator [91] to generate 9 OFDM subcarriers (A). A disinterleaver separates the odd and even subcarriers that are individually encoded with DBPSK (B) and DQPSK (C) signals, respectively. The OFDM signal is generated by combining the odd and even channels (B/C) in an optical coupler (D).

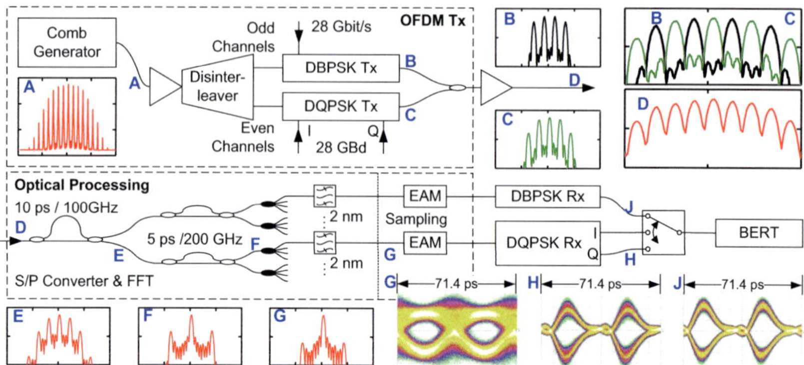

Fig. 5.2. Setup of OFDM transmission system. Two cascaded Mach-Zehnder modulators generate an optical frequency comb (A), which is split by a disinterleaver into 4 odd and 5 even channels. Spectrally adjacent subcarriers are modulated differently using decorrelated DBPSK (B) or DQPSK modulators (C), respectively. All subcarriers are combined in a coupler and transmitted (D). The OFDM nature of the signal is demonstrated by the spectral overlap of subcarriers with neighboring subcarrier sidebands, (B)/(C) and (D). The received OFDM signal (D) is processed using the "S/P converter & FFT" (D, E), where following DI stages are replaced by passive splitters (F) and optical band-pass filters. The resulting signals are sampled by electro-absorption modulators (EAM) (G) and detected using DBPSK and DQPSK receivers. Either eye diagrams (G), (H), (J), or bit error probabilities BER Fig. 5.3 were measured with a BERT. Spectra are plotted with 20 dB/div (vertically) and 2 nm/div (horizontally) in a resolution bandwidth of 0.01 nm, center of plotted spectra located at 1550 nm.

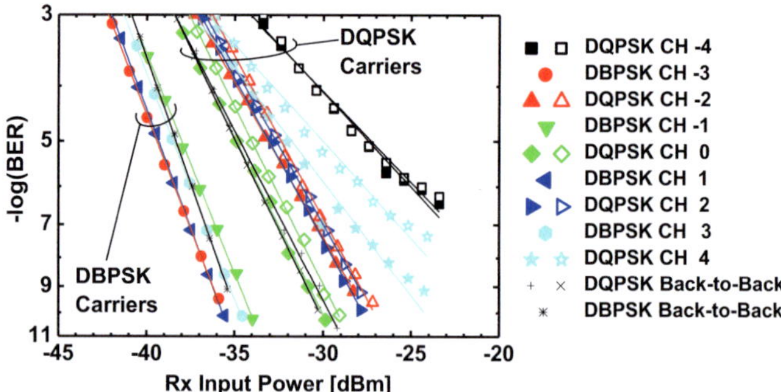

Fig. 5.3. BER performance of different subcarriers. No penalty compared to back-to-back for DBPSK carriers (−3, −1, 1, 3), no significant penalty for the central DQPSK carriers (−2, 0, 2), a 5 dB penalty or error floor for the two outer DQPSK subcarriers with 11 dB less power in the optical comb (−4, 4).

The signal with the spectrum D is really an OFDM signal, and not a dense WDM signal. First, one clearly sees from B/C and D that the subcarriers strongly overlap with neighboring subcarrier sidebands. Second, we checked the quality of reception experimentally and with simulations for many possible receiver filter shapes. A penalty-free detection is only possible with an FFT receiver. As the orthogonality of the subcarriers has to be maintained for the duration of the OFDM symbols, rise and fall times of the available transmitters have to be excluded by introducing a guard time. This results in a symbol rate that is lower than the frequency spacing of the subcarriers. An additional increase of the guard time increases the available sampling window for the receiver. The chosen guard time of 15.7 ps effectively reduces the symbol rate R to 28 GBd,

$$R = \frac{1}{T_s + T_G} \leq \frac{1}{T_s + T_{Rise/Fall}}, \quad T_s = \frac{1}{\Delta f}. \tag{5.1.3}$$

Both DBPSK as well as inphase and quadrature phase DQPSK signals were electrically decorrelated by RF delays. The resulting signals are combined to generate the 392 Gbit/s OFDM signal (D) with a symbol rate of 28 GBd. The limitations of our setup restrict us to $2^7 - 1$ long PRBS for BER measurements.

The receiver comprises the new all-optical FFT scheme followed by a preamplified receiver with differential direct-detection. The FFT processor comprises a cascade of two DIs, followed by passive splitters and band-pass filters (explained in the previous section), and the EAM sampling gates. The first DI suppresses every second subcarrier (E), the second DI every fourth subcarrier (F). The band-pass filter finally selects one of the remaining carriers (G). The inter-carrier and inter-symbol interference (see eye diagram G) is suppressed by an optical gate, which sets the FFT window. Bit error probabilities (BER) for each subcarrier have been measured (J, H).

For evaluating the FFT performance we compared the optically processed subcarriers with back-to-back (B2B) signals delivered by the DBPSK and DQPSK transmitters, respectively. The results in Fig. 5.3 show no penalty compared to the B2B performance for DBPSK and only a small penalty for the DQPSK channels. The outer channels −4 and 4 perform worse, because their subcarriers have 11 dB less power compared to the center channel.

5.1.3 Conclusion

The optical FFT can be implemented in several different configurations, all of which are based on delay interferometers. Specially by cascading delay interferometer, one could reduce the complexity and the number of DIs of the system. With the tunable delay interferometer as discussed in Section 3.1, we present the experimental demonstration of an OFDM signal reception at 392 Gbit/s. The same principle was applied in other experiments that carry data up to 26 Tbit/s [J4][J7][C7][C11][C12][C14][C17].

5.2 All Optical Wavelength Conversion

Part of the content has been published in [J9], [C24] and [C25]:

S. Sygletos, R. Bonk, T. Vallaitis, A. Marculescu, P. Vorreau, J. Li, R. Brenot, F. Lelarge, H. Duan, W. Freude, and J. Leuthold, "Filter assisted wavelength conversion with quantum-dot SOAs," J. Lightw. Technol., vol. 28, no. 6, pp. 882–897, 2010.

S. Sygletos, R. Bonk, T. Vallaitis, A. Marculescu, P. Vorreau, J. Li, R. Brenot, F. Lelarge, G. H. Duan, W. Freude, and J. Leuthold, "Optimum filtering schemes for performing wavelength conversion with QD-SOA," in Proc. of 11th Intern. Conf. on Transparent Optical Networks (ICTON'09), Ponta Delgada, Island of Sao Miguel, Portugal, vol. 1, pp. 1–4, paper Mo.C1.3, 2009 [invited].

R. Bonk, S. Sygletos, R. Brenot, T. Vallaitis, A. Marculescu, P. Vorreau, J. Li, W. Freude, F. Lelarge, G.-H. Duan, and J. Leuthold, "Optimum filter for wavelength conversion with QD-SOA," in Proc. of CLEO/IQEC, Baltimore (Maryland), USA, paper CMC6, 2009.

Wavelength conversion at high bit rates is decisive for future wavelength division multiplexing based networks [92]. Technologies for all optical wavelength conversion usually comprise nonlinear optical devices, among which semiconductor optical amplifier (SOA) is usually the choice. In particular, the combination of a single SOA followed by an optical filter has demonstrated impressive performance at bit rates up to 320 Gbit/s [93]-[96].

The principle is based on the cross gain modulation (XGM) and cross phase modulation (XPM) effects in the non-linear devices. Optical signal and a continuous wave (CW) at two different carrier frequencies are transmitted together into the non-linear device, e.g., a SOA within which the signal pulses deplete the carrier density and results in a relatively low amplification of the CW beam. Thus the information is converted onto the CW carrier, which however has an inverted pulse shape. Meanwhile, due to the change of the carrier density in the medium, the refractive index also changes thus a time varying phase shift occurs in the converted signal. With this phase information, one could utilize filter to reshape the signal into non-inverted shape.

An offset optical filter is usually the choice to shape the signal into non-inverted. However, a strong patterning effect can be observed at high speed signal because of the relatively slow carrier refilling in the SOA. A pulse reformatting optical filter is then introduced to mitigate such patterning effect [97][J10][C26][C27][C28][C31]. However, the pulse reformatting optical filter is usually not easy to fabricate and control, filters combining tunable delay interferometers [J5] and band-pass filter provides a promising solution with good flexibility [J9][C22][C24][C25][74].

Compared to bulk SOA, quantum dot (QD)-SOA provides much faster carrier refilling [98][99] and more broaden gain spectrum [100]. Therefore it's more attractive to use QD-SOA for wavelength conversion especially at high bit rates. The operating conditions and the selection of the filters have been investigated by our group [J9][C24][C25].

All optical wavelength conversion scheme using silicon-organic hybrid (SOH) waveguide in a CMOS compatible chip has also been reported [C22].

In this section we will present the principle and the experimental demonstration of a QD-SOA based wavelength conversion setup comprising tunable delay interferometers.

5.2.1 Principle of All Optical Wavelength Converter

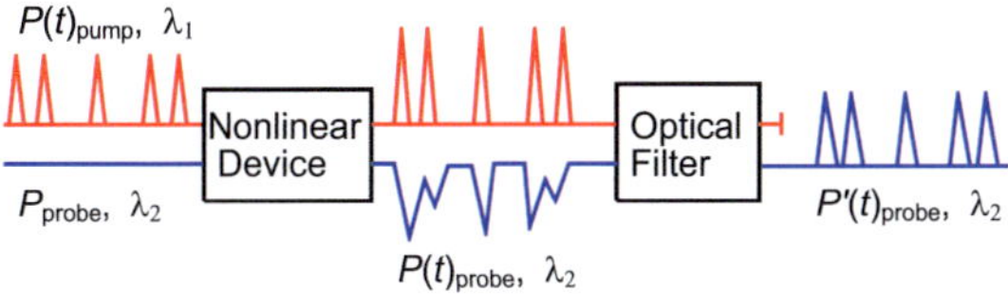

Fig. 5.4. Wavelength conversion scheme based on a nonlinear device and an optical filter. An inverted probe waveform results after the nonlinear device. The optical filter restores the initial inversion and mitigates bit patterning.

A generic configuration of the wavelength converter composed of a nonlinear device and an optical filter is illustrated in Fig. 5.4. A strong data signal at wavelength λ_1 (pump) modulates the phase and amplitude of a CW-signal (probe) entering the nonlinear device at a different wavelength λ_2. At the output of the nonlinear device this results in an inverted pulse stream, which carries the information of the input data signal. Both pump and probe wavelengths are then directed to the optical filter, which has three functions. Firstly, it blocks the pump channel, secondly it restores the non-inverted waveform of the probe channel, and thirdly it suppresses bit patterning.

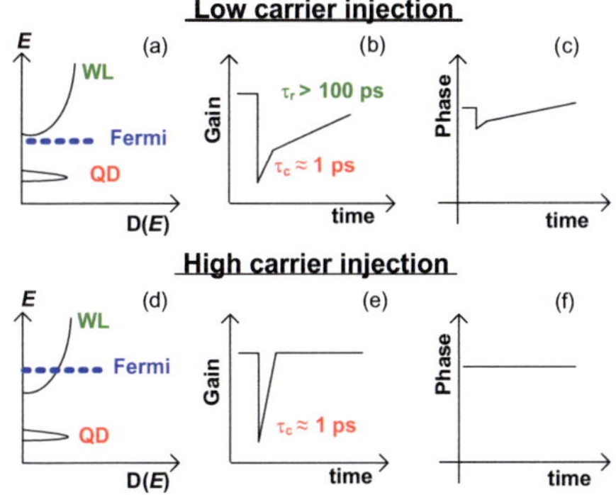

Fig. 5.5. Energy E versus density of states $D(E)$ and gain as well as phase versus time for a QD-SOA with low (upper row) and high carrier injection (lower row). In the low carrier injection regime, (a) the QD states are partially filled and the quasi-Fermi level lies at the low energy edge of the wetting layer (WL) band. (b) The gain has an initial ultrafast recovery (~ 1 ps) due to the refilling of the QD states from the WL reservoir, and afterwards the recovery slows down (~ 100 ps) by the necessary refilling of the WL states. (c) Phase effects are minor and primarily introduced by the changes of the WL carrier density. In the regime of high carrier injection, (d) the QD states are fully populated, so that the quasi-Fermi level shifts towards higher energies. (e) The gain dynamics are dominated by the ultrafast recovery of the QD states. (f) No phase modulation occurs since the WL is not depleted.

For a QD-SOA, depending on the filling of the wetting layer (WL) and QD states two operating regimes can be defined. In case of a low carrier injection the QD states are partially filled and the corresponding quasi Fermi level lays at low energy states within the conduction band, Fig. 5.5(a). Due to this low number of carriers the dynamic properties of the device are dominated by the slow refilling of the WL. Therefore, gain and phase recovery times are in the order of 100 ps [101]. This leads to strong patterning effects at high bit-rates. State-of-the-art devices so far mostly operate in this regime. Yet, in the near future we are likely to have devices that can tolerate high injection currents [102]. In such devices the quasi Fermi level will shift towards the band edges leading to a high carrier population, Fig. 5.5(b). At such a regime ultra-fast dynamics of the QD states dominate (~ 1 ps). Thus, pattern effect free operation can be expected as long as bit periods are larger than the respective QD refilling time [98]. Furthermore our calculations have shown that phase changes introduced by the spectral hole burning and carrier transition between the WL and the QD states are negligible.

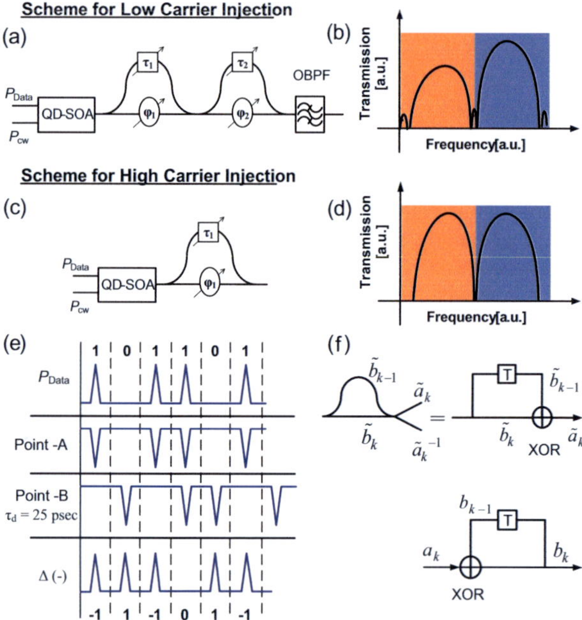

Fig. 5.6. Proposed filtering schemes for different operating regimes of QD-SOA. For the low carrier injection regime, (a) we propose a structure consisting of two cascaded DIs and an optical band-pass filter (OBPF). (b) By proper tuning its elements we may select either the blue or the red sideband of the signal. In the regime of high carrier injection, (c) we propose the scheme with a single DI. (d) It acts as a notch filter and transmits both red and blue spectral components alike. (e) The two signals in path A and B are destructively combined resulting Δ(-). (f) Logic operation performed by the optical DI, and the differential encoder.

A different optimum filter would be required at each operating regime to assist wavelength conversion. For the case of low carrier injection, the converted signal suffers from extinction ratio (ER) degradation and patterning in both amplitude and phase due to the dominance of the slow WL effects. A red and blue chirp occurs at the falling and raising curve of the inverted pulse respectively. As the energy of the red chirped component is closer to the carrier, two DIs and a bandpass filter are cascaded to suppress the red components and convert the strong blue chirped component to a non-inverted signal, see Fig. 5.6(a)(b). On the other hand, in the regime of high carrier injection, phase effects are minor and the data signal is mapped by mostly XGM onto the CW wavelength. Furthermore, due to the dominance of the fast QD dynamics, no bit patterning is present and the converted signal suffers only from a bad ER. Therefore the role of the optical filter in this case would be rather to compensate the deficiency in the ER performance. A single DI is positioned after the QD-SOA, Fig. 5.6(c)(d). The delay time in DI is chosen as 1 bit, it works similarly as a differential decoder (Fig. 5.6(e)) therefore a differential encoder need to be applied to obtain the original signals in the electrical domain, Fig. 5.6(f).

5.2.2 Experimental Results

Initially we assumed operation in the low carrier injection regime. At the input of the QD-SOA we launched a data signal consisting of a 42.7 Gbit/s RZ (33%) OOK at a wavelength of 1550 nm modulated by a pseudo-random bit sequence (PRBS) of $2^{11} - 1$. The CW (probe) channel is located at 1545 nm. The average power of the data channel was –5 dBm, and that of the CW channel it was 0 dBm. The converted signal just after the QD-SOA enters the concatenated DIs, where by setting the time delay equal to 8 ps and proper tuning of their phases the operating point maximizes the signal quality. The results of this process are illustrated in Fig. 5.7(a). The figure shows the spectrum of the modulated probe channel, obtained after the QD-SOA and after the optical filtering structure, respectively, as well as the transfer function of the filter. The signal after the SOA has a stronger blue frequency part than a red one, which accordingly favors the implementation of blue-shifted filtering. The eye diagrams of the probe channel after the QD-SOA and after the filter structure are also shown in Fig. 5.7(a). After the QD-SOA the quality of the inverted signal cannot be determined. However, after the optical filter the non-inverted waveform is characterized by a Q^2-factor of 15.6 dB and an extinction ratio of 10 dB.

Demonstration of wavelength conversion in the high carrier injection regime was not possible. Therefore, to prove the concept of our proposed filtering scheme, we performed a model experiment for the high-speed case at a much lower bit rate of 12 Gbit/s. The corresponding results are demonstrated in Fig. 5.7(b). The figure depicts the signal spectra after the QD-SOA device and after the 1-bit DI, respectively, as well as the transfer function of the DI. A pattern-effect free signal with a poor ER has been measured just after the QD-SOA. However, the 1 bit differential operation that is imposed on the signal by the DI can significantly enhance the ER. As it is shown in Fig. 5.7(b), an improved ER of 13 dB is measured. In addition, the Q^2-factor equals 16.1 dB. The output signal spectrum reflects the conversion of the format to an alternate mark inversion (AMI) [103].

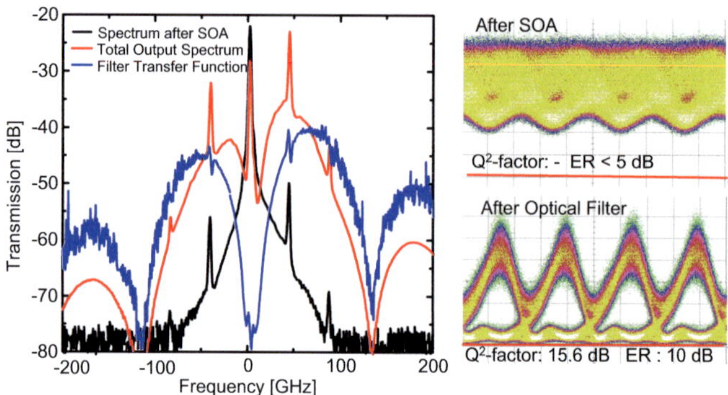

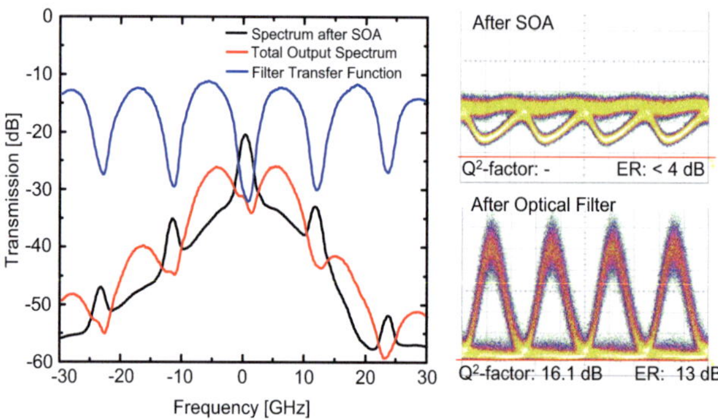

Fig. 5.7. Eye diagrams and spectra of the modulated probe channel just after the QD-SOA and at the filter output, as well as the transfer functions of the corresponding filtering schemes, (a) in the regime of low carrier injection, (b) in the regime of high carrier injection. The proposed schemes significantly improve the quality of the output signal.

5.2.3 Conclusion

We present that tunable delay interferometer can be used to implement all optical wavelength converter with high flexibility. Principle of wavelength converter using QD-SOA combined with DIs has been explained. Experimental demonstration with RZ-OOK signal up to 43 Gbit/s is shown.

The principle is furthered proved in [74] with a QD-SOA holding fast gain response. Because of the residual phase effects, the single DI is set with delay less than 1 bit. Experimental demonstration using RZ-OOK signals is performed up to 80 Gbit/s. Same principle can be also applied with a SOH waveguide, demonstration of 43 Gbit/s signal has been presented in [C22].

5.3 Chromatic Dispersion Compensation

Chromatic dispersion compensation can be achieved by using lattice filters (Fig. 5.8(a)) [84] because it has similar filter characteristics to those of finite impulse response (FIR) digital filters [104][105]. However, such filter is intricate to fabricate and characterize. A much simpler configuration has been proposed in [83] by cascading two separate delay interferometer as depicted in Fig. 5.8(b). It works as a two tap feed-forward linear equalizer. The amplitude and phases of its impulse response can be adjusted with the coupling ratios and the phase shifters in the DIs.

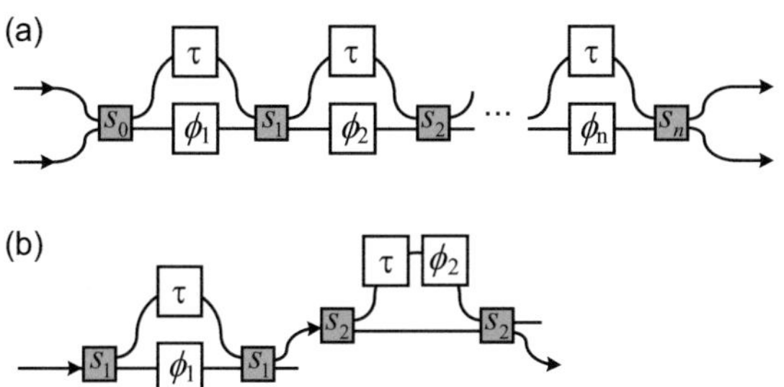

Fig. 5.8. Chromatic dispersion compensation schemes. (a) Lattice filter consists of multiple optical couplers with delay line and phase shifter in between. (b) Cascaded delay interferometers.

Given s_1, $(1-s_1)$, s_2 and $(1-s_2)$ as splitting ratios of the couplers in two DIs, the frequency response and impulse response of the cascaded DIs are as follows [83],

$$H(\omega) = \left[(1-s_1)e^{-j\omega\tau} - s_1 e^{j\phi_1} \right] \times \left[(1-s_2) - s_2 e^{j(-\omega\tau+\phi_2)} \right],$$

$$h(t) = -s_1(1-s_2)e^{j\phi_1}\delta(t) + \left[(1-s_1)(1-s_2) + s_1 s_2 e^{j(\phi_1+\phi_2)} \right] \times \delta(t-\tau) \qquad (5.3.1)$$

$$-s_2(1-s_1)e^{j\phi_2}\delta(t-2\tau).$$

From the above equation we can see that the impulse response varies at different time delays, among which the one at time τ is the strongest. Thus such cascaded DI scheme can be used to improve the signal quality for signals that suffer from inter-symbol interference (ISI) due to chromatic dispersion.

In order to illustrate the function of cascaded delay interferometer for group delay manipulating, the group delay transfer function of the cascaded delay interferometer can be derived similarly as in Eq. (3.1.4) and Eq. (3.1.5). Here we plot two examples. One is when $s_1 = 0.5$, $\phi_1 = 0°$, $\phi_2 = 0°$ and a various of $s_2 = 0.5, 0.45, 0.4, 0.35, 0.3$. The group delay versus frequency offset is plotted in Fig. 5.9(a). The second example is when $s_1 = 0.45$, $s_2 = 0.45$, $\phi_2 = 0°$, and a various of $\phi_1 = 0°, 10°, 20°, 30°, 40°$. The group delay versus frequency offset is plotted in Fig. 5.9(b). Therefore the shape of group delay transfer function depends on parameters s_1, s_2, ϕ_1, and ϕ_2. The slope of the group delay changes by varying the parameters.

Experimental demonstration of the chromatic dispersion compensation has been performed at 40 Gbit/s for NRZ signals [83].

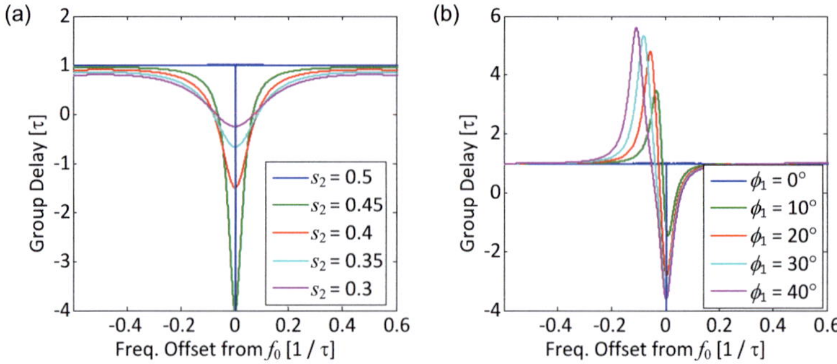

Fig. 5.9. Group delay versus frequency offset. (a) $s_1 = 0.5$, $\phi_1 = 0°$, $\phi_2 = 0°$ and a various of $s_2 = 0.5, 0.45, 0.4, 0.35, 0.3$. (b) $s_1 = 0.45$, $s_2 = 0.45$, $\phi_2 = 0°$, and a various of $\phi_1 = 0°, 10°, 20°, 30°, 40°$.

Appendix A.

A.1. Equalizer for Polarization Demultiplexing

The algorithms that used in coherent receiver systems for polarization demultiplexing have been developed for wireless communications for multiple input multiple output (MIMO) system [106][107]. A PolMUX system is a kind of MIMO system that has two inputs at orthogonal polarizations E_x and E_y which experience a state of polarization variation along the fiber channel. Two signals are mixed and form two new signals E'_x and E'_y. This process can be modeled as a Jones matrix as in Section 4.3.2.1. In the receiver the task is to estimate the channel matrix and obtain the inverse to compensate the mixing.

Given the received signals in the detector are E'_x and E'_y, a digital filter is then introduced to recover the original polarizations $\hat{E}_x$ and $\hat{E}_y$, as shown in Fig. A.1(a).

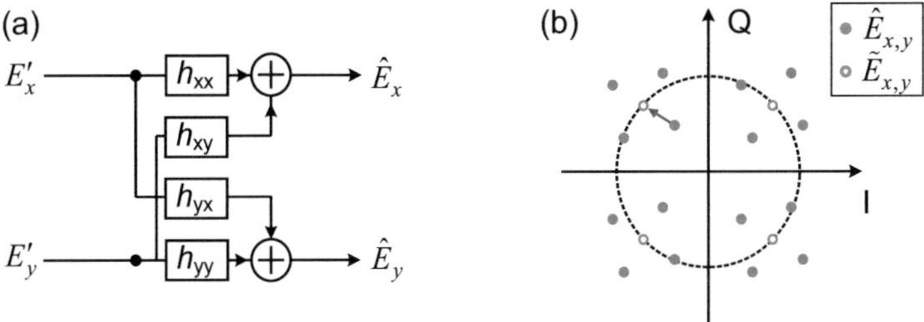

Fig. A.1. PolMUX signal equalizer. (a) Schematic drawing of an equalizer for demultiplexing polarizations. (b) Constellation diagram of a filtered signal $\hat{E}_{x,y}$ (gray solid circles) and the ideal symbols after the decision circuit (gray empty circles). The red vector indicates the difference between a detected symbol and its ideal symbol after decision circuit.

For the signals at sample n, we have [30],

$$
\begin{aligned}
\hat{E}_x(n) &= \mathbf{h}^T_{xx}(n) \cdot E'_x(n-M+1,...,n) + \mathbf{h}^T_{xy}(n) \cdot E'_y(n-M+1,...,n) \\
&= \sum_{m=1}^{M} \left[h_{xx}(n,m)\hat{E}'_x(n-m+1) + h_{xy}(n,m)\hat{E}'_y(n-m+1) \right], \\
\hat{E}_y(n) &= \mathbf{h}^T_{yx}(n) \cdot E'_x(n-M+1,...,n) + \mathbf{h}^T_{yy}(n) \cdot E'_y(n-M+1,...,n) \\
&= \sum_{m=1}^{M} \left[h_{yx}(n,m)\hat{E}'_x(n-m+1) + h_{yy}(n,m)\hat{E}'_y(n-m+1) \right],
\end{aligned}
\tag{A.1.1}
$$

where M is the tap size of the finite impulse response (FIR) filter $\mathbf{h}_{xx}$, $\mathbf{h}_{xy}$, $\mathbf{h}_{yx}$ and $\mathbf{h}_{yy}$. M can be set to be '1' if the signal at one sample is not related to its neighbors (no dispersion). In this case $[\mathbf{h}_{xx}(n), \mathbf{h}_{xy}(n); \mathbf{h}_{yx}(n), \mathbf{h}_{yy}(n)]$ is the inverse of the channel matrix as given in Eq. (4.3.2). The filter, also called equalizer is adaptive, and its coefficients are constantly updated by mainly either of the following schemes. One is the "constant modulus algorithm (CMA) and the other is the "decision directed least mean squared" (DD-LMS) algorithm. In the following we will present them separately.

Constant modulus algorithm. The equalizer attempts to minimize the variation of the sampled signal modulus based on a minimum mean square error (MMSE) principle. Assuming the signal has an average modulus of '1' as shown in Fig. A.1(b). The error quantities are given by,

$$\varepsilon_x = 1 - \left| \hat{E}_x(n) \right|^2, \quad \varepsilon_y = 1 - \left| \hat{E}_y(n) \right|^2. \tag{A.1.2}$$

Therefore we could put Eq. (A.1.1) into the above equation. Thus we get a function of ε_x and ε_y with variables of $[\mathbf{h}_{xx}, \mathbf{h}_{xy}; \mathbf{h}_{yx}, \mathbf{h}_{yy}]$. We assume that the mean squared ε_x and ε_y have a minimum where following criteria hold,

$$\frac{d\left\langle \varepsilon_x^2 \right\rangle}{d\mathbf{h}_{xx}} = 0; \quad \frac{d\left\langle \varepsilon_x^2 \right\rangle}{d\mathbf{h}_{xy}} = 0; \quad \frac{d\left\langle \varepsilon_y^2 \right\rangle}{d\mathbf{h}_{yx}} = 0; \quad \frac{d\left\langle \varepsilon_y^2 \right\rangle}{d\mathbf{h}_{yy}} = 0; \tag{A.1.3}$$

$\langle x \rangle$ denotes expectation of x.

To calculate the optimal $[\mathbf{h}_{xx}, \mathbf{h}_{xy}; \mathbf{h}_{yx}, \mathbf{h}_{yy}]$ at sample $n+1$, stochastic gradient algorithms with convergence parameter $\mu / 4$ is applied,

$$\mathbf{h}_{xx}(n+1) = \mathbf{h}_{xx}(n) - \frac{\mu}{4} \frac{d\left\langle \varepsilon_x^2 \right\rangle}{d\mathbf{h}_{xx}}, \quad \mathbf{h}_{xy}(n+1) = \mathbf{h}_{xy}(n) - \frac{\mu}{4} \frac{d\left\langle \varepsilon_x^2 \right\rangle}{d\mathbf{h}_{xy}},$$

$$\mathbf{h}_{yx}(n+1) = \mathbf{h}_{yx}(n) - \frac{\mu}{4} \frac{d\left\langle \varepsilon_y^2 \right\rangle}{d\mathbf{h}_{yx}}, \quad \mathbf{h}_{yy}(n+1) = \mathbf{h}_{yy}(n) - \frac{\mu}{4} \frac{d\left\langle \varepsilon_y^2 \right\rangle}{d\mathbf{h}_{yy}}. \tag{A.1.4}$$

The gradients are then replaced by their instantaneous values resulting in [30],

$$\mathbf{h}_{xx}(n+1) = \mathbf{h}_{xx}(n) + \mu\varepsilon_x \hat{E}_x(n) \cdot E_x'^{*}(n - M + 1, \ldots, n),$$

$$\mathbf{h}_{xy}(n+1) = \mathbf{h}_{xy}(n) + \mu\varepsilon_x \hat{E}_x(n) \cdot E_y'^{*}(n - M + 1, \ldots, n),$$

$$\mathbf{h}_{yx}(n+1) = \mathbf{h}_{yx}(n) + \mu\varepsilon_y \hat{E}_y(n) \cdot E_x'^{*}(n - M + 1, \ldots, n),$$

$$\mathbf{h}_{yy}(n+1) = \mathbf{h}_{yy}(n) + \mu\varepsilon_y \hat{E}_y(n) \cdot E_y'^{*}(n - M + 1, \ldots, n). \tag{A.1.5}$$

In order to initialize the algorithm, all tap weights are set to zero except the central of $\mathbf{h}_{xx}(n)$ and $\mathbf{h}_{yy}(n)$ which are set to unity, so that there is no coupling between the two polarizations.

Decision directed least mean square algorithm. This algorithm attempts to minimize the error vector between the filtered signal to its decision based on a minimum mean square error (MMSE) principle. The error vectors, e.g., the red line in Fig. A.1(b) can be written as,

$$\varepsilon_x(n) = \tilde{E}_x(n) - \hat{E}_x(n), \quad \varepsilon_y(n) = \tilde{E}_y(n) - \hat{E}_y(n), \tag{A.1.6}$$

where $\tilde{E}_x$ and $\tilde{E}_y$ are the ideal symbols after decision circuit. The criteria are same as for CMA. Following similar derivation, the filter is adapted by,

$$\mathbf{h}_{xx}(n+1) = \mathbf{h}_{xx}(n) + \mu\varepsilon_x \cdot E_x'^{\,*}(n-M+1,\ldots,n),$$

$$\mathbf{h}_{xy}(n+1) = \mathbf{h}_{xy}(n) + \mu\varepsilon_x \cdot E_y'^{\,*}(n-M+1,\ldots,n),$$

$$\mathbf{h}_{yx}(n+1) = \mathbf{h}_{yx}(n) + \mu\varepsilon_y \cdot E_x'^{\,*}(n-M+1,\ldots,n),$$

$$\mathbf{h}_{yy}(n+1) = \mathbf{h}_{yy}(n) + \mu\varepsilon_y \cdot E_y'^{\,*}(n-M+1,\ldots,n).$$

(A.1.7)

A.2. Theory of the Interferometer Frontend

The content of this section is a direct copy of the Journal publication [J1]:

J. Li, M. R. Billah, P. C. Schindler, M. Lauermann, S. Schuele, S. Hengsbach, U. Hollenbach, J. Mohr, C. Koos, W. Freude, and J. Leuthold, "Four-in-one interferometer for coherent and self-coherent," Opt. Express, submitted.

Part of the content of this section has been filed in a patent [P1]:

J. Li, M. Lauermann, S. Schuele, J. Leuthold, and W. Freude, "Optical detector for detecting optical signal beams, method to detect optical signals, and use of an optical detector to detect optical signals," US patent 20, 120, 224, 184, 2012.

Minor changes have been done to adjust the notations of variables.

Coherent Receiver. The operation principle of the coherent receiver is explained by using the Jones calculus. At the input we have electric fields of the signal (solid line in Fig. 3.12(e)) and LO (dashed line),

$$\vec{E}_{sig} = \begin{bmatrix} E_{sig\text{-}s} \\ E_{sig\text{-}p} \end{bmatrix} = \begin{bmatrix} A_{sig\text{-}s}e^{j\omega_{sig}t} \\ A_{sig\text{-}p}e^{j\omega_{sig}t} \end{bmatrix}, \quad \vec{E}_{LO} = \begin{bmatrix} E_{LO\text{-}s} \\ E_{LO\text{-}p} \end{bmatrix} = \begin{bmatrix} A_{LO\text{-}s}e^{j\omega_{LO}t} \\ A_{LO\text{-}p}e^{j\omega_{LO}t} \end{bmatrix}. \tag{A.2.1}$$

The signal and LO are split by two separate PBSs in two orthogonal polarizations (transmitted beam in *p*-polarization, magenta line, and reflected beam in *s*-polarization, green line). The two spatially separated beams enter the non-polarizing beam-splitter (NPBS) where they are equally split in two paths. We now focus on the respective optical paths with a particular polarization, namely *p*-polarization (magenta solid line) for the signal, and *s*-polarization (green dashed line) for the LO, point A in Fig. 3.12(e). We assume that all the components are ideal, and that there is a phase shift of '−j' between the reflected and transmitted beams after the PBS1 as well as after the NPBS. Omitting the propagator, the beam on the upper path after passing the NPBS (point B in Fig. 3.12(e)) can be written as

$$\begin{bmatrix} E_s^B \\ E_p^B \end{bmatrix} = \frac{1}{\sqrt{2}} \begin{bmatrix} 1 & 0 \\ 0 & 1 \end{bmatrix}_{\text{NPBS}} \cdot \begin{bmatrix} -jE_{LO\text{-}s} \\ E_{sig\text{-}p} \end{bmatrix} = \frac{1}{\sqrt{2}} \begin{bmatrix} -jE_{LO\text{-}s} \\ E_{sig\text{-}p} \end{bmatrix}. \tag{A.2.2}$$

This beam passes through a half waveplate (HWP). The axes of the half waveplate are at an angle of −45° with respect to the axes of the PBSs. It flips the polarization of the signal and LO to their orthogonal states. The field after the HWP at point C in Fig. 3.12(e) is

$$\begin{bmatrix} E_s^C \\ E_p^C \end{bmatrix} = \begin{bmatrix} 0 & 1 \\ 1 & 0 \end{bmatrix}_{\text{HWP--45°}} \cdot \frac{1}{\sqrt{2}} \begin{bmatrix} -jE_{LO\text{-}s} \\ E_{sig\text{-}p} \end{bmatrix} = \frac{1}{\sqrt{2}} \begin{bmatrix} E_{sig\text{-}p} \\ -jE_{LO\text{-}s} \end{bmatrix}. \qquad (A.2.3)$$

The beam is then reflected and passes through a quarter waveplate (QWP). Because magnitude and phase introduced by the two reflectors in the two paths are same, their influences cancel when the two beams interfere, so we omit the reflectors in our calculation. The axes of the QWP are aligned to the axes of the signal and LO. The QWP adds a 90° phase shift for one polarization (here, the s-polarization) so that the signal can be superimposed with an inphase LO and quadrature-phase LO separately. The field at point D in Fig. 3.12(e) is

$$\begin{bmatrix} E_s^D \\ E_p^D \end{bmatrix} = \begin{bmatrix} j & 0 \\ 0 & 1 \end{bmatrix}_{\text{QWP0°}} \cdot \frac{1}{\sqrt{2}} \begin{bmatrix} E_{sig\text{-}p} \\ -jE_{LO\text{-}s} \end{bmatrix} = \frac{1}{\sqrt{2}} \begin{bmatrix} jE_{sig\text{-}p} \\ -jE_{LO\text{-}s} \end{bmatrix}. \qquad (A.2.4)$$

The electric field on the right hand side of point E in Fig. 3.12(e) is

$$\begin{bmatrix} E_s^E \\ E_p^E \end{bmatrix} = \frac{-j}{\sqrt{2}} \begin{bmatrix} 1 & 0 \\ 0 & 1 \end{bmatrix}_{\text{NPBS}} \cdot \begin{bmatrix} -jE_{LO\text{-}s} \\ E_{sig\text{-}p} \end{bmatrix} = \frac{1}{\sqrt{2}} \begin{bmatrix} -E_{LO\text{-}s} \\ -jE_{sig\text{-}p} \end{bmatrix}. \qquad (A.2.5)$$

The beams are then reflected back onto the NPBS, where the signal and LO beams are superimposed at point F. Both the I and the Q signals interfere constructively (destructively) with the correspondingly polarized LO field. At outputs in Fig. 3.12(e),

$$\begin{bmatrix} E_{\Delta I\text{-}p} \\ E_{\Delta Q\text{-}p} \end{bmatrix} = \frac{1}{2} \begin{bmatrix} -j\left(E_{sig\text{-}p} - E_{LO\text{-}s}\right) \\ -j\left(E_{sig\text{-}p} - jE_{LO\text{-}s}\right) \end{bmatrix}, \quad \begin{bmatrix} E_{\Sigma I\text{-}p} \\ E_{\Sigma Q\text{-}p} \end{bmatrix} = \frac{1}{2} \begin{bmatrix} E_{sig\text{-}p} + E_{LO\text{-}s} \\ -\left(E_{sig\text{-}p} + jE_{LO\text{-}s}\right) \end{bmatrix}. \qquad (A.2.6)$$

The inphase and quadrature phase of the output beams are orthogonally polarized and subsequently separated by PBSs. At the four outputs the typical relations for an optical 90° hybrid are to be seen, $E_{sig\text{-}p} + E_{LO\text{-}s}$ and $E_{sig\text{-}p} - E_{LO\text{-}s}$ for the inphase component, and $E_{sig\text{-}p} + jE_{LO\text{-}s}$ and $E_{sig\text{-}p} - jE_{LO\text{-}s}$ for the quadrature-phase component.

The relation for the s-polarization of the signal and p-polarization of LO can be deduced in the same way. Assuming the mirror cube in the left corner will introduce a phase shift π between s and p-polarizations, the resulting field at the outputs in Fig. 3.12(e) are

$$\begin{bmatrix} E_{\Sigma I\text{-}s} \\ E_{\Delta Q\text{-}s} \end{bmatrix} = \frac{1}{2} \begin{bmatrix} -j\left(E_{sig\text{-}s} + E_{LO\text{-}p}\right) \\ E_{sig\text{-}s} - jE_{LO\text{-}p} \end{bmatrix}, \quad \begin{bmatrix} E_{\Delta I\text{-}s} \\ E_{\Sigma Q\text{-}s} \end{bmatrix} = \frac{1}{2} \begin{bmatrix} -\left(E_{sig\text{-}s} - E_{LO\text{-}p}\right) \\ j\left(E_{sig\text{-}s} + jE_{LO\text{-}p}\right) \end{bmatrix}. \qquad (A.2.7)$$

Thus in this polarization and phase diverse coherent receiver two optical 90° hybrids are nested in one physical device.

Self-coherent receiver. The input has two orthogonal polarizations,

$$\vec{E}_{sig} = \begin{bmatrix} E_{sig\text{-}s} \\ E_{sig\text{-}p} \end{bmatrix}. \qquad (A.2.8)$$

The signal is split at PBS1, and both polarizations are further split at NPBS in two paths. The upper path has a time delay τ with respect to the path on the right hand side. After the NPBS, both beams pass a quarter waveplate (QWP). The axes of the QWPs (in white) are at a −45° angle with respect to the optical axes of the PBSs, so that they convert the linear polarizations (*s* and *p*) into circular polarizations. Taking for example the *s*-polarization (blue line), we find at point A, (the symbol '*' denotes the convolution operator)

$$
\begin{bmatrix} E_s^A \\ E_p^A \end{bmatrix} = \delta(t-\tau) * \left\{ \frac{1}{2}\begin{bmatrix} 1+j & 1-j \\ 1-j & 1+j \end{bmatrix}_{QWP-45°} \cdot \frac{1}{\sqrt{2}}\begin{bmatrix} 1 & 0 \\ 0 & 1 \end{bmatrix}_{NPBS} \cdot \begin{bmatrix} jE_{sig\text{-}s}(t) \\ 0 \end{bmatrix} \right\}
$$

$$
= \frac{E_{sig\text{-}s}(t-\tau)}{2\sqrt{2}}\begin{bmatrix} -1+j \\ 1+j \end{bmatrix}.
$$

(A.2.9)

In order to superimpose the signal with its I and Q delayed copy, another quarter waveplate (gray, see Fig. 3.12(j)) at an angle of 0° to the optical axes of PBSs intercepts the upper path. It introduces a relative 90° phase shift between the two polarizations in the same beam. Omitting the common propagator for the two paths, we have at point B in Fig. 3.12(j)

$$
\begin{bmatrix} E_s^B \\ E_p^B \end{bmatrix} = \begin{bmatrix} j & 0 \\ 0 & 1 \end{bmatrix}_{QWP0°} \cdot \frac{E_{sig\text{-}s}(t-\tau)}{2\sqrt{2}}\begin{bmatrix} -1+j \\ 1+j \end{bmatrix} = \frac{E_{sig\text{-}s}(t-\tau)}{2\sqrt{2}}\begin{bmatrix} -1-j \\ 1+j \end{bmatrix}.
$$

(A.2.10)

The beam on the right path goes through a QWP under −45°as well. At point C we find

$$
\begin{bmatrix} E_s^C \\ E_p^C \end{bmatrix} = \frac{1}{2}\begin{bmatrix} 1+j & 1-j \\ 1-j & 1+j \end{bmatrix}_{QWP-45°} \cdot \frac{-j}{\sqrt{2}}\begin{bmatrix} 1 & 0 \\ 0 & 1 \end{bmatrix}_{NPBS} \cdot \begin{bmatrix} jE_{sig\text{-}s}(t) \\ 0 \end{bmatrix} = \frac{E_{sig\text{-}s}(t)}{2\sqrt{2}}\begin{bmatrix} 1+j \\ 1-j \end{bmatrix}.
$$

(A.2.11)

Therefore in each beam the powers of the new *s* and *p*-polarizations (referred to the PBSs axes) are equal. Birefringent elements other than QWP can also be used as long as the powers of the new *s* and *p*-polarized fields are equal.

The delayed and undelayed beams are then superimposed in the NPBS. The electric fields at the outputs are

$$
\begin{bmatrix} E_{\Delta Q\text{-}s}(t) \\ E_{\Delta I\text{-}s}(t) \end{bmatrix} = \frac{e^{-j135°}}{2\sqrt{2}}\begin{bmatrix} E_{sig\text{-}s}(t) - jE_{sig\text{-}s}(t-\tau) \\ E_{sig\text{-}s}(t) - E_{sig\text{-}s}(t-\tau) \end{bmatrix},
$$

$$
\begin{bmatrix} E_{\Sigma Q\text{-}s}(t) \\ E_{\Sigma I\text{-}s}(t) \end{bmatrix} = \frac{e^{-j45°}}{2\sqrt{2}}\begin{bmatrix} E_{sig\text{-}s}(t) + jE_{sig\text{-}s}(t-\tau) \\ E_{sig\text{-}s}(t) + E_{sig\text{-}s}(t-\tau) \end{bmatrix}.
$$

(A.2.12)

Therefore the output PBSs separate the two polarizations to retrieve the I and Q components of the input signal at *s*-polarization.

For the *p*-polarization of the input signal (magenta line), we measure its I and Q components at the outputs as indicated

$$
\begin{bmatrix} E_{\Delta Q\text{-}p}(t) \\ E_{\Delta I\text{-}p}(t) \end{bmatrix} = \frac{e^{j45°}}{2\sqrt{2}}\begin{bmatrix} E_{sig\text{-}p}(t) - jE_{sig\text{-}p}(t-\tau) \\ -\left(E_{sig\text{-}p}(t) - E_{sig\text{-}p}(t-\tau)\right) \end{bmatrix},
$$

$$
\begin{bmatrix} E_{\Sigma Q\text{-}p}(t) \\ E_{\Sigma I\text{-}p}(t) \end{bmatrix} = \frac{e^{j135°}}{2\sqrt{2}}\begin{bmatrix} E_{sig\text{-}p}(t) + jE_{sig\text{-}p}(t-\tau) \\ -\left(E_{sig\text{-}p}(t) + E_{sig\text{-}p}(t-\tau)\right) \end{bmatrix}.
$$

(A.2.13)

Thus, in self-coherent receiver, four delay interferometers (I and Q for each of the two polarizations *s* and *p*) are spatially folded into one single delay interferometer and share the same time delay and phase shifter.

With the tunable delay and the adjustable liquid crystals, the receiver frontend can be easily switched between coherent and self-coherent receptions without replacing any component. Naturally, for coherent reception a LO has to be added.

A.3. Signal Processing of *E(t)* by Delay Interferometers

The content of this section is a direct copy from the Journal publication [J2]:
J. Li, R. Schmogrow, D. Hillerkuss, P. Schindler, M. Nazarathy, C. Schmidt-Langhorst, S.-B. Ezra, C. Koos, W. Freude, and J. Leuthold, "A self-coherent receiver for PolMUX-signals," Opt. Express, vol. 20, no. 19, pp. 21413-21433, 2012.
Minor changes have been done to adjust the notations of variables.

In this section we derive the mathematical operation performed by two DIs (Fig. 4.8(b)) onto a signal sequence $E(t)$.

The self-coherent receiver basically comprises two DIs with photo detectors at the constructive and destructive output ports. The operations performed by a single DI onto a signal $E(t)$ therefore given at the constructive and destructive ports are

$$E_{out1}(t) = E(t) + E(t-\tau),$$
$$E_{out2}(t) = E(t) - E(t-\tau). \tag{A.3.1}$$

In the photodiodes the fields are converted into photo currents that are proportional to the square of the field magnitude,

$$I_{out1}(t) \propto |E(t) + E(t-\tau)|^2$$
$$\propto |E(t)|^2 + |E(t-\tau)|^2 + 2|E(t)||E(t-\tau)|\cos(\angle E(t) - \angle E(t-\tau)),$$
$$I_{out2}(t) \propto |E(t) - E(t-\tau)|^2 \tag{A.3.2}$$
$$\propto |E(t)|^2 + |E(t-\tau)|^2 - 2|E(t)||E(t-\tau)|\cos(\angle E(t) - \angle E(t-\tau)),$$

where $\angle x$ denotes the phase of a complex quantity x. After the balanced photodiode receiver, the differential output current is

$$I_{BR}(t) = I_{out1}(t) - I_{out2}(t) \propto |E(t)||E(t-\tau)|\cos(\angle E(t) - \angle E(t-\tau)). \tag{A.3.3}$$

This operation is performed for the inphase (I in Fig. 4.8(b)) as well as for the quadrature phase DI (Q in Fig. 4.8(b)). Because of the $\pi/2$ phase offset, the corresponding photocurrents $I_{BR}^{(I)}$ and $I_{BR}^{(Q)}$ are

$$I_{BR}^{(I)}(t) \propto |E(t)||E(t-\tau)|\cos(\angle E(t) - \angle E(t-\tau)),$$

$$I_{BR}^{(Q)}(t) \propto |E(t)||E(t-\tau)|\cos(\angle E(t) - \angle E(t-\tau) + \frac{\pi}{2}) \tag{A.3.4}$$

$$\propto -|E(t)||E(t-\tau)|\sin(\angle E(t) - \angle E(t-\tau)).$$

After digital acquisition we combine the I and Q channels in form of a complex signal,

$$u(t) = I_{BR}^{(I)}(t) - j\, I_{BR}^{(Q)}(t)$$

$$\propto |E(t)||E(t-\tau)|\cos(\angle E(t) - \angle E(t-\tau)) + j|E(t)||E(t-\tau)|\sin(\angle E(t) - \angle E(t-\tau)) \quad (A.3.5)$$

$$\propto |E(t)||E(t-\tau)|\exp\big[j(\angle E(t) - \angle E(t-\tau))\big] = E(t)E^*(t-\tau).$$

As the inphase and quadrature phase signals are normalized in the DSP section, we simplify the notation by writing $u(t) = E(t)E^*(t-\tau)$. Given the signal is sampled at discrete times $t_n = n\tau$, we can also write this equation as $u(n) = E(n)E^*(n-1)$.

A.4. Channel Estimation with Training Sequences

The content of this section is a direct copy from the Journal publication [J2]:
J. Li, R. Schmogrow, D. Hillerkuss, P. Schindler, M. Nazarathy, C. Schmidt-Langhorst, S.-B. Ezra, C. Koos, W. Freude, and J. Leuthold, "A self-coherent receiver for PolMUX-signals," Opt. Express, vol. 20, no. 19, pp. 21413-21433, 2012.
Minor changes have been done to adjust the notations of variables.

The training sequence for the two polarizations is chosen to be

$$\begin{array}{c} A_x(1...8):...0 \;\; 1 \;\; 1 \;\big|\; 1 \;\; 1 \;\; 0 \;\; 0 \;\big|\; A_x(8)... \\ \hline A_y(1...8):...0 \;\; 1 \;\; 1 \;\big|\; 0 \;\; 0 \;\; 1 \;\; 1 \;\big|\; A_y(8)... \end{array}, \qquad (A.4.1)$$

where $A_x(8)$ and $A_y(8)$ can be any symbol from the transmitted constellation. The preamble with a starting zero serves as a uniquely identifiable symbol sequence. For convenience we set $E_{x,y} = A_{x,y}$, i. e., we omit encoding the symbols onto an optical carrier which does not change our channel estimation process. At the receiver, we have

$$\begin{bmatrix} E'_x(n) \\ E'_y(n) \end{bmatrix}_{n=1...8} = \begin{bmatrix} C_{11} & C_{12} \\ C_{21} & C_{22} \end{bmatrix}\begin{bmatrix} E_x(n) \\ E_y(n) \end{bmatrix}_{n=1...8}$$

$$= \left\{ \begin{bmatrix} 0 \\ 0 \end{bmatrix}, \begin{bmatrix} C_{11}+C_{12} \\ C_{21}+C_{22} \end{bmatrix}, \begin{bmatrix} C_{11}+C_{12} \\ C_{21}+C_{22} \end{bmatrix}, \begin{bmatrix} C_{11} \\ C_{21} \end{bmatrix}, \begin{bmatrix} C_{11} \\ C_{21} \end{bmatrix}, \begin{bmatrix} C_{12} \\ C_{22} \end{bmatrix}, \begin{bmatrix} C_{12} \\ C_{22} \end{bmatrix}, \begin{bmatrix} C_{11}E_x(8)+C_{12}E_y(8) \\ C_{21}E_x(8)+C_{22}E_y(8) \end{bmatrix} \right\}. \qquad (A.4.2)$$

After reception with the DI and digital acquisition we measure

$$\begin{bmatrix} u'_x(n) \\ u'_y(n) \end{bmatrix}_{n=1...8} = \begin{bmatrix} E'^{*}_x(n-1)E'_x(n) \\ E'^{*}_y(n-1)E'_y(n) \end{bmatrix}_{n=1...8}$$

$$= \left\{ \begin{bmatrix} 0 \\ 0 \end{bmatrix}, \begin{bmatrix} 0 \\ 0 \end{bmatrix}, \begin{bmatrix} (C_{11}+C_{12})^*(C_{11}+C_{12}) \\ (C_{21}+C_{22})^*(C_{21}+C_{22}) \end{bmatrix}, \begin{bmatrix} (C_{11}+C_{12})^*C_{11} \\ (C_{21}+C_{22})^*C_{21} \end{bmatrix}, \begin{bmatrix} C_{11}^*C_{11} \\ C_{21}^*C_{21} \end{bmatrix}, \begin{bmatrix} C_{11}^*C_{12} \\ C_{21}^*C_{22} \end{bmatrix}, \begin{bmatrix} C_{12}^*C_{12} \\ C_{22}^*C_{22} \end{bmatrix}, \begin{bmatrix} C_{12}^*(C_{11}E_x(8)+C_{12}E_y(8)) \\ C_{22}^*(C_{21}E_x(8)+C_{22}E_y(8)) \end{bmatrix} \right\}. \qquad (A.4.3)$$

From the elements in Eq. (A.4.3) we use the first four elements as starting delimiter. Also we do not use $u'_{x,y}(8)$ for channel estimation. This reduces the number of equations to 6. This should be sufficient equations to derive the complex channel elements if the absolute value and the absolute phase are not needed,

$$u'_x(5) = C_{11}{}^*C_{11} = |C_{11}|^2, \qquad u'_y(5) = C_{21}{}^*C_{21} = |C_{21}|^2,$$
$$u'_x(6) = C_{11}{}^*C_{12}, \qquad u'_y(6) = C_{21}{}^*C_{22}, \qquad (A.4.4)$$
$$u'_x(7) = C_{12}{}^*C_{12} = |C_{12}|^2, \qquad u'_y(7) = C_{22}{}^*C_{22} = |C_{22}|^2.$$

We employ the same notation as in Eq. (4.3.4) for the phase factor $\breve{C}$ of a complex quantity C, and the estimation $\hat{\mathbf{C}}$ of the channel matrix $\mathbf{C}$ results from Eq. (A.4.4),

$$\hat{C}_{11} = |u'_x(5)|^{\frac{1}{2}} \exp\left(-j\angle u'_x(6)\right) = C_{11} \exp\left[j(-\angle C_{12})\right] = C_{11}\breve{C}_{12}{}^*,$$
$$\hat{C}_{12} = |u'_x(7)|^{\frac{1}{2}} = C_{12} \exp\left[j(-\angle C_{12})\right] = C_{12}\breve{C}_{12}{}^*,$$
$$\hat{C}_{21} = |u'_y(5)|^{\frac{1}{2}} \exp\left(-j\angle u'_y(6)\right) = C_{21} \exp\left[j(-\angle C_{22})\right] = C_{21}\breve{C}_{22}{}^*, \qquad (A.4.5)$$
$$\hat{C}_{22} = |u'_y(7)|^{\frac{1}{2}} = C_{22} \exp\left[j(-\angle C_{22})\right] = C_{22}\breve{C}_{22}{}^*.$$

Compared to the channel matrix $\mathbf{C}$, its estimation $\hat{\mathbf{C}}$ has common phase factors $\breve{C}_{12}{}^*$ and $\breve{C}_{22}{}^*$ for its upper and lower rows, respectively.

As a test, we use this estimated channel matrix to recover the transmitted symbols as discussed in Section 4.3.2.4. We first define our starting symbol $\hat{E}'_x(8)$ and $\hat{E}'_y(8)$,

$$\begin{bmatrix} \hat{E}'_x(8) \\ \hat{E}'_y(8) \end{bmatrix} = \begin{bmatrix} \hat{C}_{11}E_x(8) + \hat{C}_{12}E_y(8) \\ \hat{C}_{21}E_x(8) + \hat{C}_{22}E_y(8) \end{bmatrix} = \begin{bmatrix} E'_x(8)\breve{C}_{12}{}^* \\ E'_y(8)\breve{C}_{22}{}^* \end{bmatrix}. \qquad (A.4.6)$$

Substituting this into Eq. (4.3.8), we find $\hat{E}'_x(9)$ and $\hat{E}'_y(9)$,

$$\begin{bmatrix} \hat{E}'_x(9) \\ \hat{E}'_y(9) \end{bmatrix} = \begin{bmatrix} u'_x(9)/\hat{E}'_x{}^*(8) \\ u'_y(9)/\hat{E}'_y{}^*(8) \end{bmatrix} = \begin{bmatrix} E'_x(9)\breve{C}_{12}{}^* \\ E'_y(9)\breve{C}_{22}{}^* \end{bmatrix}. \qquad (A.4.7)$$

Then we multiply this column matrix with the inverse of the estimated channel matrix $\hat{\mathbf{C}}$,

$$\begin{bmatrix} \hat{E}_x(9) \\ \hat{E}_y(9) \end{bmatrix} = \begin{bmatrix} \hat{C}_{11} & \hat{C}_{12} \\ \hat{C}_{21} & \hat{C}_{22} \end{bmatrix}^{-1} \begin{bmatrix} \hat{E}'_x(9) \\ \hat{E}'_y(9) \end{bmatrix} = \frac{1}{\hat{C}_{11}\hat{C}_{22} - \hat{C}_{12}\hat{C}_{21}} \begin{bmatrix} \hat{C}_{22} & -\hat{C}_{12} \\ -\hat{C}_{21} & \hat{C}_{11} \end{bmatrix} \begin{bmatrix} E'_x(9)\breve{C}_{12}{}^* \\ E'_y(9)\breve{C}_{22}{}^* \end{bmatrix}$$
$$= \frac{1}{C_{11}C_{22}\breve{C}_{12}{}^*\breve{C}_{22}{}^* - \hat{C}_{12}\hat{C}_{21}\breve{C}_{12}{}^*\breve{C}_{22}{}^*} \qquad (A.4.8)$$
$$\begin{bmatrix} C_{22}\breve{C}_{12}{}^*\breve{C}_{22}{}^* & -C_{12}\breve{C}_{12}{}^*\breve{C}_{22}{}^* \\ -C_{21}\breve{C}_{12}{}^*\breve{C}_{22}{}^* & C_{11}\breve{C}_{12}{}^*\breve{C}_{22}{}^* \end{bmatrix} \begin{bmatrix} E'_x(9) \\ E'_y(9) \end{bmatrix}.$$

After re-arranging terms we find,

$$\begin{bmatrix} \hat{E}_x(9) \\ \hat{E}_y(9) \end{bmatrix} = \frac{1}{C_{11}C_{22} - C_{12}C_{21}} \begin{bmatrix} C_{22} & -C_{12} \\ -C_{21} & C_{11} \end{bmatrix} \begin{bmatrix} E'_x(9) \\ E'_y(9) \end{bmatrix}$$

$$= \begin{bmatrix} C_{11} & C_{12} \\ C_{21} & C_{22} \end{bmatrix}^{-1} \begin{bmatrix} C_{11} & C_{12} \\ C_{21} & C_{22} \end{bmatrix} \begin{bmatrix} E_x(9) \\ E_y(9) \end{bmatrix} = \begin{bmatrix} E_x(9) \\ E_y(9) \end{bmatrix}. \qquad (A.4.9)$$

Thus, the symbols for $n = 9$ can be correctly recovered. After the decision circuit (which is introduced because of practical reasons), $\tilde{E}_x(9)$ and $\tilde{E}_y(9)$ are multiplied with the estimated channel matrix to generate the complex symbol reference for the next symbol recovery,

$$\begin{bmatrix} \hat{E}'_x(9) \\ \hat{E}'_y(9) \end{bmatrix} = \begin{bmatrix} \hat{C}_{11} & \hat{C}_{12} \\ \hat{C}_{21} & \hat{C}_{22} \end{bmatrix} \begin{bmatrix} \tilde{E}_x(9) \\ \tilde{E}_y(9) \end{bmatrix} = \begin{bmatrix} \hat{C}_{11}E_x(9) + \hat{C}_{12}E_y(9) \\ \hat{C}_{21}E_x(9) + \hat{C}_{22}E_y(9) \end{bmatrix}$$

$$= \begin{bmatrix} \breve{C}_{12}^{*}(C_{11}E_x(9) + C_{12}E_y(9)) \\ \breve{C}_{22}^{*}(C_{21}E_x(9) + C_{22}E_y(9)) \end{bmatrix} = \begin{bmatrix} E'_x(9)\breve{C}_{12}^{*} \\ E'_y(9)\breve{C}_{22}^{*} \end{bmatrix}. \qquad (A.4.10)$$

Repeating the procedure Eq. (A.4.7) - (A.4.10), we can reconstruct the subsequent symbols.

There are two extreme cases that cannot be treated with Eq. (A.4.4), namely if either of the $u'_{x,y}(5...7)$ is close to zero, i. e., if there is no polarization crosstalk $C_{12}, C_{21} \rightarrow 0$, or if the polarization states are interchanged $C_{11}, C_{22} \rightarrow 0$,

$$\begin{bmatrix} C_{11} & C_{12} \\ C_{21} & C_{22} \end{bmatrix} = \begin{bmatrix} 1 & 0 \\ 0 & 1 \end{bmatrix} \quad \text{or} \quad \begin{bmatrix} C_{11} & C_{12} \\ C_{21} & C_{22} \end{bmatrix} = \begin{bmatrix} 0 & 1 \\ 1 & 0 \end{bmatrix}. \qquad (A.4.11)$$

In this case, $u'_{x,y}(6)$ in Eq. (A.4.3) becomes very small, and its argument needed in Eq. (A.4.5) for calculating $\hat{C}$ is very inaccurate. Depending on $u'_{x,y}(5...7)$, the starting fields $E'_x(8)$ and $E'_y(8)$ are equal to the transmitted fields $E_x(8)$ and $E_y(8)$, or they are equal to the cross-polarized transmitted fields $E_y(8)$ and $E_x(8)$. The subsequent fields $E_{x,y}(9)$, $E_{x,y}(10)$, ... can then be recovered as described previously.

A.5. Transmitter for a Self–Coherent System

In Section 4.3, we introduce a self-coherent system transmitting a PolMUX-DQPSK signal comprising a normal DQPSK and a phase progressive DQPSK. To perform the experiment, an emulator using a single modulator is presented to demonstrate the principle. In practice, two polarizations need to be modulated separately. The transmitter for this system can be implemented in two different ways.

The first method is using a software-defined transmitter, where digital-analog converters (DCA) with high resolution bits are required [42]. As shown in Fig. A.2, the laser source is split and fed into two separate IQ modulators. Each is driven by a software-defined electrical signal source. One modulator is modulated with normal DQPSK signal, the other is modulated with a 45° phase progressive DQPSK signal. The two signals are then combined in a polarization beam combiner (PBC) generating a PolMUX signal.

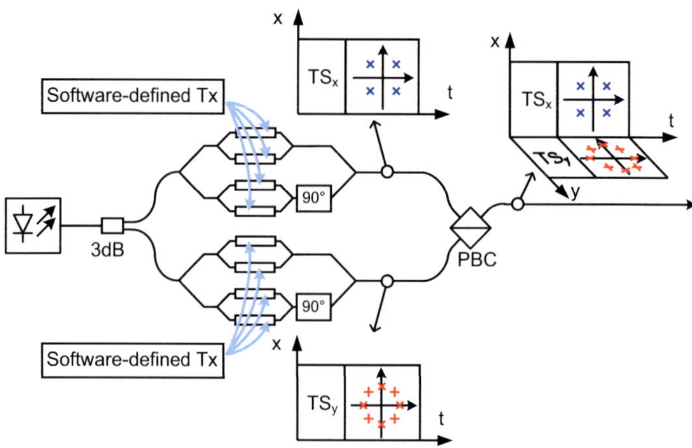

Fig. A.2. Schematic drawing of a transmitter setup where PolMUX signal is generated by two separated IQ modulators. Each modulator is driven by a software-defined electrical source. One generates normal DQPSK, and the other generates a phase progressive DQPSK. Two signals are combined in a PBC resulting a PolMUX signal.

The second method uses normal binary electrical signals. Therefore no software-defined transmitter is needed. However, an additional phase modulator is required. The phase modulator is positioned behind one of the IQ modulator as shown in Fig. A.3. It is driven by a ½ clock signal to create a 45° phase shift between each adjacent symbols, so that the differential phase is '45°', '135°', '−135°' and '−45°' instead of '0°', '90°', '180°' and '−90°'. The training sequence need be adjusted accordingly to compensate the phase shift introduced by the phase modulator.

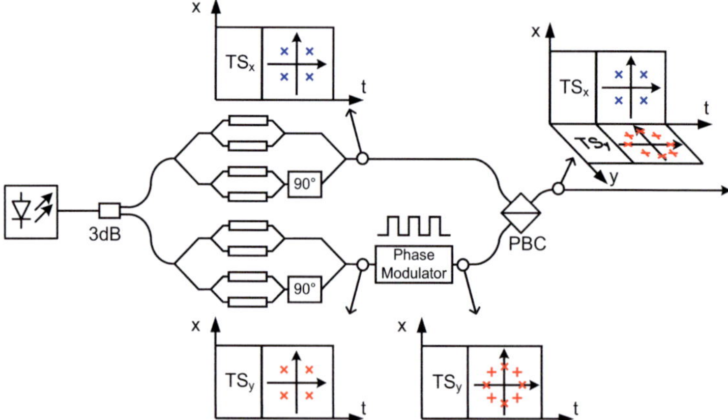

Fig. A.3. Schematic drawing of a PolMUX transmitter using normal binary electrical signals. It utilizes a clocked phase modulator to generate the progressive phase DQPSK signal.

Glossary

Calligraphic Symbols

$\mathcal{E}_{\text{in},1}(f)$	Fourier transform of $E_{\text{in},1,2}(t)$
$\mathcal{E}_{\text{out},1,2}(f)$	Fourier transform of $E_{\text{out},1,2}(t)$
$\mathcal{M}_i$	Characteristic matrix of one optical thin film, Eq. (2.3.14)

Greek Symbols

ε	Electric permittivity of the medium, Eq. (2.2.5)
ε_0	Electric permittivity of vacuum, $\varepsilon_0 = 8.854\,188\times10^{-12}\,\text{As/(Vm)}$, Eq. (2.3.2)
$\varepsilon_{i,r,t}$	Relative electric permittivity of the incidence, reflecting, and transmitting medium, Eq. (2.3.3)
$\varepsilon_{x,y}$	Error quantities of equalizer, Eq. (A.1.2)
η	Quantum efficiency, Eq. (2.2.1)
Θ	Beam divergence of a Gaussian beam, unit rad
$\theta_{1,2,3\ldots}$	Angle of incidence reflection and transmission at the optical interface
θ_c	The offset angle between two coordination systems, Eq. (2.4.7)
$\theta_{i,r,t}$	Angle of incidence, reflection and transmission, Eq. (2.3.1)
θ_n	Offset angle between the eigenstates of the PBC/PBS and the optical axes of fiber, Eq. (4.3.2)
λ	Wavelength, unit m
μ	Magnetic permeability of the medium, Eq. (2.2.5)
μ	Step size of an equalizer, Eq. (A.1.5)
μ_0	Magnetic permeability of vacuum, $\mu_0 = 1.256\,637\times10^{-6}\,\text{Vs/(Am)}$, Eq. (2.3.2)
$\mu_{i,r,t}$	Relative magnetic permeability of the incidence, reflecting and transmitting medium, Eq. (2.3.4)
ν_{FM}	Modulation frequency of the pilot tone, Eq. (3.1.9)
$\delta\nu_d$	Frequency deviation $\delta\nu_d$ of the pilot tone, Eq. (3.1.9)
τ	Time delay in delay interferometer, unit s
τ_{DGDn}	Differential group delay, Eq. (4.3.2)
$\delta\tau$	Change of the delay time τ in the DI, Eq. (3.1.8)
$\Upsilon_{s,p}$	Magnetic field and electric field relation factor, Eq. (2.3.12)
$\Phi_{1,2}(f)$	Phase response of the delay interferometer, Eq. (3.1.4)
ϕ_{12}^I, ϕ_{21}^I	Phase shift from one port to the other in a coupler 'I', unit rad, Eq. (3.1.1)

$\phi_{12}^{II}, \phi_{21}^{II}$	Phase shift from one port to the other in a coupler 'II', unit rad, Eq. (3.1.1)
$\phi_{I,Q}$	Phase offset between two arms in a DI, unit rad
$\Delta\phi_{\mathrm{pol}}$	Phase retardation introduced by the liquid crystal, unit rad
$\delta\phi_{\mathrm{PMDn}}$	Phase offset between two orthogonal polarizations, unit rad, Eq. (4.3.2)
φ_{0sig}	Phase of the signal complex amplitude, unit rad, Eq. (2.2.7)
$\varphi_{0x,y}$	Optical phase of signal on x and y polarization, unit rad, Eq. (2.1.1)
$\varphi_{x,y}^{\varepsilon}$	Phase error, Eq. (4.2.1)
$\Delta\varphi_{\mathrm{pol}}$	Phase retardation of the signal, unit rad
$\Delta\varphi_{t,r}$	Phase retardation of the transmitted and reflected beams, Eq. (2.3.6)
ω	Angular frequency, unit rad / s, Eq. (2.2.4)
$\omega_{0x,y}$	Carrier angular frequency at polarization x and y, unit rad / s, Eq. (2.1.1)
ω_{LO}	Carrier angular frequency of local oscillator, unit rad / s, Eq. (2.2.9)
ω_{sig}	Carrier angular frequency of the signal, unit rad / s, Eq. (A.2.1)
$\Delta\omega$	Signal bandwidth, Eq. (4.3.2)

Latin Symbols

$A_{0x,y}$	Amplitude of the signal at polarization x or y, unit V / m, Eq. (2.1.1)
A_{eff}	Effective area of the beam, unit m^2, Eq. (2.2.3)
$A_{x,y}$	Complex amplitude of an optical signal at polarization x or y
$\breve{A}_x$	$\breve{A}_x = A_x / |A_x|$, Eq. (4.3.4)
$a_{1,2}$	Power loss factor at the upper and lower arms of DI, Eq. (3.1.1)
$a_{I,II}$	Power loss factor at coupler 'I' and 'II' of DI, Eq. (3.1.1)
B	Bandwidth of baseband signal
C	Channel matrix, Eq. (4.3.2)
C_{DI}	Complexity factor of an optical FFT circuit based on cascading DIs
C_{ij}	Element of the channel matrix, Eq. (4.3.2)
C_{std}	Complexity factor of conventional optical FFT circuit
c	Speed of light in vacuum, $c = 1/\sqrt{\varepsilon_0\mu_0} = 299\,792\,458$ m/s
D_i	Data encoded on each subcarrier of an OFDM signal, Eq. (2.1.2)
d_i	Thickness of optical thin film
d_w	Working distance of an optical collimator, unit m
DGD	Differential group delay, Eq. (3.1.7)
$\vec{E}$	Vector electric field, unit V/m, Eq. (2.1.1)
$E_{i,r,t}$	Electric field of incident, reflected and transmitted light
$E_{\mathrm{in},1,2}$	Electric field of input signal at port 1 or 2, Eq. (3.1.1)
$\vec{E}_{LO}$	Vector electric field of local oscillator, unit V/m, Eq. (A.2.1)
$E_{LO-s,p}$	Electric field of LO for s or p-polarization, unit V/m, Eq. (A.2.1)
$E_{LOx,y}$	Electric field of LO for polarization x or y, unit V/m, Eq. (2.2.7)

$E_{\text{out},1,2}$	Electric field of output signal at port 1 or 2, Eq. (3.1.1)
$E'_{r_{sII},pII}$	Electric field of reflected beam for s light and p light from boundary 'II' arrives at boundary 'I', Eq. (2.3.7)
$E_{s,p}$	Electric field of s light or p light, unit V/m
$E_{s,p,I,II}$	Electric field of s or p light at boundary 'I' and 'II', Eq. (2.3.7)
$\vec{E}_{sig}$	Vector electric field of optical signal, unit V/m, Eq. (A.2.1)
$E_{sig\text{-}s,p}$	Electric field of signal for s or p-polarization, unit V/m, Eq. (A.2.1)
$E_{sigx,y}$	Electric field of signal for polarization x or y, unit V/m, Eq. (2.2.7)
$E_{x,y}$	Electric field, unit V/m, Eq. (2.1.1)
$E'_{x,y}$	Electric field after polarization change via fiber channel, Eq. (4.3.2)
$\hat{E}_{x,y}$	Estimated electric field at polarization x or y, unit V/m, Eq. (4.3.8)
$\hat{E}'_{x,y}$	Estimated electric field after polarization change via fiber channel
$\hat{\hat{E}}_{x,y}$	Estimated electric field of the signal at polarization x or y. The magnitude is normalized to '1', Eq. (4.3.9)
$\tilde{E}_{x,y}$	Electric field at polarization x or y after the decision circuits
e	Elementary charge, $e = 1.60217646 \times 10^{-19}$ C
$\angle E_{sig}$	Phase of optical signal, unit rad, Eq. (2.2.7)
$\angle E_{LO}$	Phase of optical local oscillator, unit rad, Eq. (2.2.11)
ER	Extinction ratio
f	Frequency, unit Hz
f_0	Center carrier frequency of an OFDM signal, unit Hz, Eq. (2.1.2); Center Frequency where destructive interference occurs at one output of DI, while in the other output constructive interference occurs
f_c	Central carrier frequency of the signal
f_i	Subcarrier center frequency of an OFDM signal, unit Hz, Eq. (2.1.4)
f_m	Subcarrier center frequency of an OFDM signal, unit Hz
f_n	Subcarrier center frequency of an OFDM signal, unit Hz
f_p	Central carrier frequency of the pilot tone
Δf	Frequency spacing between subcarriers of an OFDM signal, unit Hz, Eq. (2.1.2)
Δf_c	The frequency offset between carrier center frequency of the signal and DI, Eq. (3.1.9)
Δf_p	The frequency offset between pilot tone and DI, Eq. (3.1.9)
FSR	Free spectral range
$\vec{H}$	Vector magnetic field, unit A/m, Eq. (2.2.5)
$H_{1,2}(f)$	Power transfer function for output 1 or 2, Eq. (3.1.3)
$H_{s,p}(f)$	Power transfer function for s or p polarization, Eq. (3.1.7)
$H_{sI,II}$	Magnetic field of s light at boundary 'I' and 'II', Eq. (2.3.7)
h	Planck constant, 6.626068×10^{-34} m^2 kg / s
h_i	Optical distance between two boundaries, Eq. (2.3.10)
$h_{xx,xy,yx,yy}$	FIR filter coefficient, Eq. (A.1.1)
I_{ac}	Output current of the phase monitoring photodiode, Eq. (3.1.9)

I_{BR}	Flow of current at the output of the balanced detector, unit A, Eq. (2.2.9)
$I_{i,r,t}$	Intensity of the incident, reflected, and transmitted beam
I_I	Flow of current at the inphase channel, unit A, Eq. (2.2.11)
I_Q	Flow of current at the quadrature phase channel, unit A, Eq. (2.2.11)
$I_{out1,2}$	Flow of current of the photo diode at the outputs, unit A, Eq. (2.2.7)
I_p	Flow of current, unit A, Eq. (2.2.1)
I_{sig}	Optical field intensity of the signal, unit W/m^2, Eq. (2.2.3)
i	Subcarrier index of an OFDM signal, Eq. (2.1.2)
IF	Intermediate frequency
$J_{1,2}$	Bessel functions of order 1 and 2, Eq. (3.1.9)
k	Propagation vector
L	Layer index of a multi-layer optical thin film system, Eq. (2.3.16)
M	Number of frequency spacing between two subcarriers of an OFDM signal, Eq. (2.1.3); Tap size of FIR filter, Eq. (A.1.1)
$\mathbf{M}_{r,t}$	Transfer matrix of optical interface for the reflected and transmitted beams, Eq. (2.4.3)
m	Sample index, Eq. (5.1.1)
m_{ij}	Element of characteristic matrix of an optical thin film, Eq. (2.3.16)
m_{max}	Constructive interference fringes number, Eq. (3.1.8)
m_{min}	Destructive interference fringes number, Eq. (3.1.8)
N	Number of subcarriers of an OFDM signal, Eq. (2.1.2); Sample number, Eq. (5.1.1)
n	Sample index, Eq. (5.1.1); Refractive index of the material, Eq. (2.2.5)
$n_{H,L,0,s}$	Refractive index of the coating material with higher refractive index, material with lower refractive index, the incidence medium, and the medium of substrate, Eq. (2.3.28)
$n_{i,r,t}$	Refractive index of the incident, reflecting and transmitting medium, Eq. (2.3.1)
$\mathbf{P}$	Jones matrix describing phase offset after PBS, Eq. (4.3.2)
P_{in}	Incident optical power, unit W, Eq. (2.2.1)
$PDFS$	Polarization dependent frequency shift
PDL	Polarization dependent loss, Eq. (3.1.7)
$\mathbf{PMD}_n$	Jones matrix describing PMD, Eq. (4.3.2)
R	Symbol rate, Eq. (5.1.3)
$\mathbf{R}_{1\text{-}2}$	Transfer matrix of coordination system rotation, Eq. (2.4.9)
$\mathbf{R}_n$	Jones matrix describing polarization rotation, Eq. (2.4.7) (4.3.2)
$R_{s,p}$	Reflectance for s and p lights, Eq. (2.3.3)
R_{xcorr}	Cross correlation between two subcarriers of an OFDM signal, Eq. (2.1.3)

$r_{s,p}$	Reflection coefficient for s and p lights, Eq. (2.3.1)
$\angle r_{s,p}$	Phase shift of the reflected beam for s and p lights, Eq. (2.3.6)
$s(t)$, $S(f)$	Symbol of the signal, Eq. (2.1.2)
$\hat{s}$	Estimated symbol of the signal
$\mathbf{S_{I,II}}$	Matrix of coupler I and II, Eq. (3.1.1)
$s_{I,II}$	Splitting ratio of coupler I and II, Eq. (3.1.1)
$s_{1,2}$	Splitting ratio of couplers in DI 1 or 2, Eq. (5.3.1)
SpS	Samples per symbol
$\mathbf{T}$	Matrix of time delay in delay interferometer, Eq. (3.1.1)
T	Time duration, unit s, Eq. (2.2.5)
T_G	Guard time, Eq. (5.1.3)
$T_{s,p}$	Transmittance for s and p lights, Eq. (2.3.4)
$T_{Rise/Fall}$	The rise and fall times of the transmitter, Eq. (5.1.3)
T_s	Symbol period, unit s
t	Time, unit s
t_0	Starting time, unit s
$t_{g1,2}(f)$	Group delay response of the delay interferometer, Eq. (3.1.5)
$t_{g-s,p}(f)$	Group delay response of the delay interferometer for s or p polarization, Eq. (3.1.7)
$t_{s,p}$	Transmission coefficient for s or p light, Eq. (2.3.1)
$\angle t_{s,p}$	Phase shift of the transmitted beam for s or p light, Eq. (2.3.6)
$u_{x,y}$	Complex amplitude of the combined signal, Eq. (2.2.12)
$u'_{x,y}$	Complex signal after IQ-DI for polarization x and y with arbitrary polarization change
$\hat{u}_{x,y}$	Estimated complex signal after IQ-DI for polarization x and y
V	Electric voltage, unit v
$v_{i,r,t}$	Phase velocity of light in the incident, reflecting and transmitting medium
W	Depletion-region width of a PIN diode, unit m
w	Weighting factor
w	Beam radius, unit m, Eq. (2.4.1)
w_0	Beam waist, unit m, Eq. (2.4.1)
X_{E_m}, X_{O_m}	The even and odd DFT of size $N/2$ for even and odd inputs, Eq. (5.1.2)
X_m	The DFT results of x_n, Eq. (5.1.1)
x	(Vertical / s light) spatial coordinate, Eq. (2.1.1)
x_n	Time samples, Eq. (5.1.1)
y	(Horizontal / p light) spatial coordinate, Eq. (2.1.1)
z	Spatial coordinate in the direction of beam propagation, Eq. (2.4.1)
z_r	Rayleigh range, unit m, Eq. (2.4.1)

Acronyms

8PSK	8-phase shift keying
ADC	Analog-to-digital converter
AMI	Alternate mark inversion
AR	Anti-reflective
ASE	Amplified spontaneous emission
ASK	Amplitude-shift keying
B2B	Back-to-back
BER	Bit error ratios
BERT	Bit error ratio tester
BPSK	Binary phase shift keying
CD	Chromatic dispersion
CMA	Constant modulus algorithm
CMOS	Complementary metal–oxide–semiconductor
CSRZ	Carrier suppressed return to zero
CW	Continuous wave
CWDM	Coarse WDM
D8PSK	Differential 8-phase shift keying
DAC	Digital-analog converters
DBPSK	Differential binary phase shift keying
DCA	Digital communication analyser
DD	Decision directed
DFT	Discrete Fourier transform
DGD	Differential group delay
DI	Delay interferometer
DP	Differentially precoded
DPSK	Differential phase shift keying
DQPSK	Differential quadrature phase shift keying
DSP	Digital signal processing
DWDM	Dense WDM
EAM	Electro-absorption modulators
EDEC	Electronic demodulator error compensation
EDFA	Erbium doped fiber amplifier
ER	Extinction ratio
EuroFOS	European Commission's Network of Excellence
EVM	Error vector magnitude
FFT	Fast Fourier transforms
FIR	Finite impulse response
FSK	Frequency-shift keying
FSR	Free spectral range

HWP	Half waveplate
I	Inphase
IDFT	Inverse discrete Fourier transform
IF	Intermediate frequency
IFFT	Inverse fast Fourier transform
IP	Internet Protocol
ISF	Israeli Science Foundation
ISI	Inter-symbol interference
ITU	International Telecommunication Union
KIT	Karlsruhe Institute of Technology
KSOP	Karlsruhe School of Optics and Photonics
LC	Liquid crystal
LIGA	X-ray lithography, electroplating (galvanic), and molding (Abformung)
LMS	Least mean square
LO	Local oscillator
MIMO	Multiple-input and multiple-output
MMI	Multimode interference coupler
MMSE	Minimum mean square error
MSPE	Multi symbol phase estimation
MZI	Mach-Zehnder interferometer
NPBS	Non-polarizing beam splitter
NRZ	Non-return to zero
OBPF	Optical band-pass filter
OFDM	Optical orthogonal division multiplexing
OFE	Optical frontend
OOK	On-off keying (format)
OSA	Optical spectrum analyzer
OSNR	Optical signal to noise ratio
PBC	Polarization beam combiner
PBS	Polarization beam splitter
PDFS	Polarization dependent frequency shift
PDL	Polarization dependent loss
PFR	Polarization and field recovery
PLC	Planar lightwave circuit
PLL	Phase locked loop
PM	Polarization maintaining
PMD	Polarization mode dispersion
PolDEMUX	Polarization demultiplexing
PolMUX	Polarization multiplexing
PolSK	Polarization shift keying
PRBS	Pseudorandom binary sequence

PSK	Phase-shift keying
Q	Quadrature phase
QAM	Quadrature amplitude modulation
QD	Quantum dot
QPSK	Quadrature phase shift keying
QWP	Quarter waveplates
RF	Radio frequency
Rx	Receiver
RZ	Return-to-zero
SCD	Self-coherent detection
SE	Spectral efficiency
SOA	Semiconductor optical amplifier
SOH	Silicon-organic hybrid
SOP	State of polarization
S/P	Serial-to-parallel
SpS	Samples per symbol
TEM	Transverse electromagnetic
TS	Training sequence
Tx	Transmitter
WDM	Wavelength division multiplexing
WL	Wetting layer
XGM	Cross gain modulation
XOR	Exclusive or
XPM	Cross phase modulation

References

[1] Cisco Visual Networking Index: Forecast and Methodology, 2011-2016, 2012.

[2] Cisco Visual Networking Index: Global Mobile Data Traffic Forecast Update, 2011-2016, 2012.

[3] S. Bucci, S. Gastaldello, and L. Razzetti, "The 400 G Photonic Service Engine," Alcatel-Lucent white paper, 2012.

[4] R.A. Griffin, R.I. Johnstone, R.G. Walker, J. Hall, S.D. Wadsworth, K. Berry, A.C. Carter, M.J. Wale, J. Hughes, P.A. Jerram, and N.J. Parsons, "10 Gb/s optical differential quadrature phase shift key (DQPSK) transmission using GaAs/AlGaAs integration," in Proc. of Optical Fiber Commun. Conf., postdeadline paper FD6, 2002.

[5] D. van den Borne, S.L. Jansen, E. Gottwald, P.M. Krummrich, G.D. Khoe, and H. de Waardt, "1.6-b/s/Hz spectrally efficient transmission over 1700 km of SSMF using 40 × 85.6-Gb/s POLMUX-RZ-DQPSK," J. Lightw. Technol. 25(1), 222–232, 2007.

[6] A.H. Gnauck, G. Raybon, S. Chandrasekhar, J. Leuthold, C. Doerr, L. Stulz, A. Agarwal, S. Banerjee, D. Grosz, S. Hunsche, A. Kung, A. Marhelyuk, D. Maywar, M. Movassaghi, X. Liu, C. Xu, X. Wei, and D.M. Gill, "2.5 Tb/s (64×42.7 Gb/s) transmission over 40×100 km NZDSF using RZ-DPSK format and all-Raman-amplified spans," in Proc. of Optical Fiber Commun. Conf., postdeadline paper FC2, 2002.

[7] A. Gnauck, G. Charlet, P. Tran, P. Winzer, C. Doerr, J. Centanni, E. Burrows, T. Kawanishi, T. Sakamoto, and K. Higuma, "25.6-Tb/s WDM transmission of polarization-multiplexed RZ-DQPSK signals," J. Lightw. Technol. 26(1), 79–84, 2008.

[8] M. Yagi, S. Satomi, and S. Ryu, "Field trial of 160-Gb/s, polarization-division multiplexed RZ-DQPSK transmission system using automatic polarization control," in in Proc. of Optical Fiber Commun. Conf., paper OTuT7, 2008.

[9] B. Koch, R. Noé, V. Mirvoda, D. Sandel, V. Filsinger, and K. Puntsri, "40-krad/s polarization tracking in 200-Gb/s PDM-RZ-DQPSK transmission over 430 km," IEEE Photon. Technol. Lett. 22(9), 613–615, 2010.

[10] C.R. Doerr, and L. Chen, "Monolithic PDM-DQPSK receiver in silicon," in Proc. of European Conf. on Optical Commun., postdeadline paper PD3.6, 2010.

[11] S. Chandrasekhar, X. Liu, A. Konczykowska, F. Jorge, J. Dupuy, and J. Godin, "Direct detection of 107-Gb/s polarization-multiplexed RZ-DQPSK without optical polarization demultiplexing," IEEE Photon. Technol. Lett. 20(22), 1878–1880, 2008.

[12] R. Nagarajan, J. Rahn, M. Kato, J. Pleumeekers, D. Lambert, V. Lal, H.-S. Tsai, A. Nilsson, A. Dentai, M. Kuntz, R. Malendevich, J. Tang, J. Zhang, T. Butrie, M. Raburn, B. Little, W. Chen, G. Goldfarb, V. Dominic, B. Taylor, M. Reffle, F. Kish,

and D. Welch, "10 Channel, 45.6 Gb/s per channel, polarization-multiplexed DQPSK, InP receiver photonic integrated circuit," J. Lightw. Technol. **29**(4), 386–395, 2011.

[13] J. Salz, "Digital transmission over cross-coupled linear channels," AT&T Tech. J. **64**, 2153, 1985.

[14] J. Salz, "Modulation and detection for coherent lightwave communications," IEEE Commun. Mag. **24**(6), 38, 1986.

[15] E. Basch, and T. Brown, "Introduction to coherent optical fiber transmission," IEEE Commun. Mag. **23**(5), 23–30, 1985.

[16] T. Okoshi, "Recent advances in coherent optical fiber communication systems," J. Lightw. Technol. **5**(1), 44–52, 1987.

[17] T. Kimura, "Coherent optical fiber transmission," J. Lightw. Technol. **5**(4), 414–428, 1987.

[18] T. Okoshi, and K. Kikuchi, "Coherent optical fiber communications," Kluwer Academic, Boston, 1988.

[19] R.A. Linke, and A.H. Gnauck, "High-capacity coherent lightwave systems," J. Lightw. Technol. **6**(11), 1750–1769, 1988.

[20] J.R. Barry, "Performance of coherent optical receivers," in Proc. of IEEE **78**(8), 1369–1394, 1990.

[21] P.S. Henry, and S.D. Persoinick, "Coherent lightwave communications," IEEE press, Piscataway, NJ, 1990.

[22] S. Betti, G. de Marchis, and E. Iannone, "Coherent optical communication systems," Wiley, New York, 1995.

[23] S. Ryu, "Coherent lightwave communication systems," Artec House, Boston, 1995.

[24] J.M. Kahn and K.-P. Ho, "Spectral efficiency limits and modulation/detection techniques for DWDM systems," IEEE J. Selected Topics Quantum Electron. **10**(2), 259–272, 2004.

[25] T. Okoshi, and Y.H. Cheng, "Four-port homodyne receiver for optical fibre communications comprising phase and polarisation diversities," Electron. Lett. **23**(8), 377–378, 1987.

[26] C.R.S. Fludger, T. Duthel, D. Van den Borne, C. Schulien, E.D. Schmidt, T. Wuth, J. Geyer, E. De Man, G.D. Khoe, and H. de Waardt, "Coherent equalization and POLMUX-RZ-DQPSK for robust 100-GE transmission," J. Lightw. Technol. **26**(1), 64–72, 2008.

[27] D.-S. Ly-Gagnon, S. Tsukamoto, K. Katoh, and K. Kikuchi, "Coherent detection of optical quadrature phase-shift keying signals with carrier phase estimation," J. Lightw. Technol. **24**(1), 12–21, 2006.

[28] S. Okamoto, K. Toyoda, T. Omiya, K. Kasai, M. Yoshida, and M. Nakazawa, "512QAM (54 Gbit/s) coherent optical transmission over 150 km with an optical bandwidth of 4.1 GHz," in Proc. of European Conf. on Optical Commun., postdeadline paper PD2.3, 2010.

[29] Y. Koizumi, K. Toyoda, M. Yoshida, and M. Nakazawa, "1024QAM (60 Gbit/s) single-carrier coherent optical transmission over 150 km," Opt. Express **20**(11), 12508-12514, 2012.

[30] S. Savory, "Digital filters for coherent optical receivers," Opt. Express **16**(2), 804–817, 2008.

[31] N. Kikuchi, K. Mandai, S. Sasaki, and K. Sekine, "Proposal and first experimental demonstration of digital incoherent optical field detector for chromatic dispersion compensation," in Proc. of European Conf. on Optical Commun., paper Th444, 2006.

[32] X. Liu, S. Chandrasekhar, A.H. Gnauck, C.R. Doerr, I. Kang, D. Kilper, L.L. Buhl, and J. Centanni, "DSP-enabled compensation of demodulator phase error and sensitivity improvement in direct-detection 40-Gb/s DQPSK," in Proc. of European Conf. on Optical Commun., postdeadline paper Th4.4.5, 2006.

[33] X. Liu, S. Chandrasekhar, and A. Leven, "Digital self-coherent detection," Opt. Express **16**(2), 792–803, 2008.

[34] N. Kikuchi, and S. Sasaki, "Highly Sensitive Optical Multilevel Transmission of Arbitrary Quadrature-Amplitude Modulation (QAM) Signals With Direct Detection," J. Lightw. Technol. **28**(1), 123–130, 2010.

[35] L. Christen, Y.K. Lize, S. Nuccio, L. Paraschis, and A.E. Willner, "Experimental demonstration of reduced complexity 43-Gb/s RZ-DQPSK rate-tunable receiver," IEEE Photon. Technol. Lett. **20**(13), 1166–1168, 2008.

[36] C.R. Doerr, D.M. Gill, A.H. Gnauck, L.L. Buhl, P.J. Winzer, M.A. Cappuzzo, A. Wong-Foy, E.Y. Chen, and L.T. Gomez, "Monolithic demodulator for 40-Gbs DQPSK using a star coupler," J. Lightw. Technol. **24**(1), 171–174, 2006.

[37] L. Zimmermann, K. Voigt, G. Winzer, K. Petermann, and C.M. Weinert, "C-band optical 90° hybrids based on Silicon-on insulator 4×4 waveguide coupler," IEEE Photon. Technol. Lett. **21**(3), 143–145, 2009.

[38] W. Shieh, H. Bao and Y. Tang, "Coherent optical OFDM: theory and design," Opt. Express **16**(2), 841–859, 2008.

[39] J. Sakaguchi, Y. Awaji, N. Wada, A. Kanno, T. Kawanishi, T. Hayashi, T. Taru, T. Kobayashi, and M. Watanabe, "109-Tb/s (7×97×172-Gb/s SDM/WDM/PDM) QPSK transmission through 16.8-km homogeneous multi-core fiber," in Proc. of Optical Fiber Commun. Conf., postdeadline paper PDPB6, 2011.

[40]   P.J. Winzer, "Optical transmitters, receivers, and noise," in Wiley Encyclopedia of Telecommunications, 2002.

[41]   X. Zhou, and J.J. Yu, "Multi-level, multi-dimensional coding for high-speed and high-spectral-efficiency optical transmission," J. Lightw. Technol. **27**(16), 3641–3653, 2009.

[42]   R. Schmogrow, D. Hillerkuss, M. Dreschmann, M. Huebner, M. Winter, J. Meyer, B. Nebendahl, C. Koos, J. Becker, W. Freude, and J. Leuthold, "Real-time software-defined multiformat transmitter generating 64QAM at 28 GBd," IEEE Photon. Technol. Lett. **22**(21), 1601–1603, 2010.

[43]   Cisco white paper, "Fiber types in Gigabit optical communications", 2008.

[44]   S. Jansen, "OFDM for Optical Communications," in Proc. of Optical Fiber Commun. Conf., SC341, 2012.

[45]   A.H. Gnauck, P.J. Winzer, L.L. Buhl, T. Kawanishi, T. Sakamoto, M. Izutsu, and K. Higuma, "12.3-Tb/s C-Band DQPSK transmission at 3.2 b/s/ Hz spectral efficiency," in Proc. of European Conf. on Optical Commun., paper Th4.1.2, 2006.

[46]   A.H. Gnauck, G. Charlet, P. Tran, P. Winzer, C. Doerr, J. Centanni, E. Burrows, T. Kawanishi, T. Sakamoto, and K. Higuma, "25.6-Tb/s C+L-band transmission of polarization multiplexed RZ-DQPSK signals," in Proc. of Optical Fiber Commun. Conf., postdeadline paper PDP19, 2007.

[47]   G.P. Agrawal, "Fiber-optic communication systems," Wiley Series, 1997.

[48]   N. Sigron, I. Tselniker, and M. Nazarathy, "Carrier phase estimation for optically coherent QPSK based on Wiener-optimal and adaptive Multi-Symbol Delay Detection (MSDD)," Opt. Express **20**(3), 1981–2003, 2012.

[49]   C. Kim, and G. Li, "Direct-detection optical differential 8-level phase-shift keying (OD8PSK) for spectrally efficient transmission," Opt. Express **12**(15), 3415–3421, 2004.

[50]   G. Bosco, and P. Poggiolini, "On the joint effect of receiver impairments on direct-detection DQPSK systems," J. Lightw. Technol. **24**(3), 1323–1333, 2006.

[51]   F. Derr, "Coherent optical QPSK intradyne system: concept and digital receiver realization," J. Lightw. Technol. **10**(9), 1290–1296, 1992.

[52]   R. Noe, "Phase noise-tolerant synchronous QPSK/BPSK baseband-type intradyne receiver concept with feedforward carrier recovery," J. Lightw. Technol. **23**(2), 802–808, 2005.

[53]   A. Leven, N. Kaneda, U.-V. Koc, and Y.-K. Chen, "Frequency estimation in intradyne reception," IEEE Photon. Technol. Lett. **19**(6), 366–368, 2007.

[54]   D. van den Borne, S. Jansen, G. Khoe, H. de Wardt, S. Calabro, and E. Gottwald, "Differential quadrature phase shift keying with close to homodyne performance

based on multi-symbol phase estimation," in Proc. of IEE Seminar on Optical Fiber Comm. and Electronic Signal Processing, ref. No. 2005–11310, 2005.

[55] E. Hecht, "Optics" 4. ed., Addison-Wesley, 2002.

[56] http://en.wikipedia.org/wiki/Gaussian_beam

[57] R.M.A. Azzam, and H.K. Khanfar, "Polarization properties of retroreflecting right-angle prisms," Appl. Opt. **47**(3), 359-364, 2008.

[58] B.C. Park, T.B. Eom, and M.S. Chung, "Polarization properties of cube-corner retroreflectors and their effects on signal strength and nonlinearity in heterodyne interferometers," Appl. Opt. **35**(22), 4372–4380, 1996.

[59] http://en.wikipedia.org/wiki/Waveplate

[60] J. Leuthold, B. Mikkelsen, G. Raybon, C.H. Joyner, J.L. Pleumeekers, B.I. Miller, K. Dreyer, and R. Behringer, "All-optical wavelength conversion between 10 and 100 Gb/s with SOA delayed-interference configuration," Opt. Quantum Electron. **33**(7/10), 939–952, 2001.

[61] B. Mikkelsen, C. Rasmussen, P. Mamyshev, and F. Liu, "Partial DPSK with excellent filter tolerance and OSNR sensitivity," Electron. Lett. **42**(23), 1363–1364, 2006.

[62] Y. Nasu, K. Hattori, T. Saida, Y. Hashizume, and Y. Sakamaki, "Silica-based adaptive-delay DPSK demodulator with a cascaded Mach-Zehnder interferometer configuration," in Proc. of European Conf. on Optical Commun., paper We.8.E.5, 2010.

[63] H. Kawakami, E. Yoshida, Y. Miyamoto, and M. Oguma, "Analysing the penalty induced by PD lambda of MZI in DQPSK receiver using novel measuring technique," Electron. Lett. **43**(2), 121–122, 2007.

[64] H. Kawakami, E. Yoshida, Y. Miyamoto, M. Oguma, and T. Itoh, "Simple phase offset monitoring technique for 43 Gbit/s optical DQPSK receiver," Electron. Lett. **44**(6), 437–438, 2008.

[65] Z. Tao, A. Isomura, T. Hoshida, and J.C. Rasmussen, "Dither-free, accurate, and robust phase offset monitor and control method for optical DQPSK demodulator," in Proc. of European Conf. on Optical Commun., paper Mo.3.D.5, 2007.

[66] L. Rovati, U. Minoni, M. Bonardi, and F. Docchio, "Absolute distance measurement using comb-spectrum interferometry," J. of Optics **29**, 121–127, 1998.

[67] A. Cabral, and J. Rebordao, "Accuracy of frequency-sweeping interferometry for absolute distance metrology," Opt. Eng., SPIE **46**(7), 073602, 2007.

[68] D. Caplan, M. Stevens, and J. Carney, "High-sensitivity multi-channel single-interferometer DPSK receiver," Opt. Express **14**(23), 10984–10989, 2006.

[69] A.A. Freschi, and J. Frejlich, "Adjustable phase-control in stabilized interferometry," Opt. Lett. **20**(6), 635–637, 1995.

[70] C.R. Doerr, P.J. Winzer, Y.-K. Chen, S. Chandrasekhar, M.S. Rasras, L. Chen, T.-Y. Liow, K.-W. Ang, and G.-Q. Lo, "Monolithic polarization and phase diversity coherent receiver in silicon," J. Lightw. Technol. **28**(4), 520–525, 2010.

[71] J. Rahn, G. Goldfarb, H.-S. Tsai, W. Chen, S. Chu, B. Little, J. Hryniewicz, F. Johnson, W. Chen, T. Butrie, J. Zhang, M. Ziari, J. Tang, A. Nilsson, S. Grubb, I. Lyubomirsky, J. Stewart, R. Nagarajan, F. Kish, and D.F. Welch, "Low-power, polarization tracked 45.6 GB/s per wavelength PM-DQPSK receiver in a 10-channel integrated module," in Proc. of Optical Fiber Commun. Conf., paper OThE2, 2010.

[72] C.R. Doerr, and L. Chen, "Monolithic PDM-DQPSK receiver in silicon," in Proc. of European Conf. on Optical Commun., postdeadline paper PD3.6, 2010.

[73] Y.C. Hsieh, "Free-space optical hybrid," US Patent Application 02223932 A1, 2007.

[74] C. Meuer, C. Schmidt-Langhorst, R. Bonk, H. Schmeckebier, D. Arsenijević, G. Fiol, A. Galperin, J. Leuthold, C. Schubert, and D. Bimberg, "80 Gb/s wavelength conversion using a quantum-dot semiconductor optical amplifier and optical filtering," Opt. Express **19**(6), 5134–5142, 2011.

[75] W. Menz, and J. Mohr, "Mikrosystemtechnik für Ingenieure," Zweite erweitert Auflage, VCH Verlag, 1997.

[76] N. Kikuchi, K. Mandai, K. Sekine, and S. Sasaki, "Incoherent 32-level optical multilevel signaling technologies," J. Lightw. Technol. **26**(1), 150–157, 2008.

[77] J. Zhao, M.E. McCarthy, and A.D. Ellis, "Electronic dispersion compensation using full optical field reconstruction in 10 Gbit/s OOK based systems," Opt. Express **16**(20), 15353–15365, 2008.

[78] R. Noe, "PLL-free synchronous QPSK polarization multiplex / diversity receiver concept with digital I&Q baseband processing," IEEE Photon. Technol. Lett. **17**(4), 887–889, 2005.

[79] X. Wei, A.H. Gnauck, D.M. Gill, X. Liu, U.-V. Koc, S. Chandrasekhar, G. Raybon, and J. Leuthold, "Optical pi/2-DPSK and its tolerance to filtering and polarization-mode dispersion," IEEE Photon. Technol. Lett. **15**(11), 1639–1641, 2003.

[80] M. Oerder, and H. Meyr, "Digital filter and square timing recovery," IEEE Trans. Commun. **36**(5), 605–612, 1988.

[81] R. Schmogrow, B. Nebendahl, M. Winter, A. Josten, D. Hillerkuss, S. Koenig, J. Meyer, M. Dreschmann, M. Huebner, C. Koos, J. Becker, W. Freude, and J. Leuthold, "Error vector magnitude as a performance measure for advanced modulation formats," IEEE Photon. Technol. Lett. **24**(1), 61–63, 2012.

[82] A.J. Viterbi, and A.N. Viterbi, "Nonlinear estimation of PSK-modulated carrier phase with application to burst digital transmission," IEEE Trans. Inf. Theory, **29**(4), 543–551, 1983.

[83] C.R. Doerr, S. Chandrasekhar, P.J. Winzer, A.R. Chraplyvy, A.H. Gnauck, L.W. Stulz, R. Pafchek, E. Burrows, "Simple multichannel optical equalizer mitigating intersymbol interference for 40-Gb/s nonreturn-to-zero signals," J. Lightw. Technol. **22**(1), 249–256, 2004.

[84] K. Jinguji and M. Kawachi, "Synthesis of coherent two-port lattice-form optical delay-line circuit," J. Lightw. Technol. **13**(1), 73–82, 1995.

[85] D.J. Geisler, N.K. Fontaine, R.P. Scott, T. He, K. Okamoto, J.P. Heritage, and S.J.B. Yoo, "3 b/s/Hz 1.2 Tb/s packet generation using optical arbitrary waveform generation based optical transmitter," in Proc. of Optical Fiber Commun. Conf., paper JThA29, 2009.

[86] H. Sanjoh, E. Yamada, and Y. Yoshikuni, "Optical orthogonal frequency division multiplexing using frequency/time domain filtering for high spectral efficiency up to 1 bit/s/Hz," in Proc. of Optical Fiber Commun. Conf., paper ThD1, 2009.

[87] J.W. Cooley, and J.W. Tukey, "An algorithm for the machine calculation of complex Fourier series," Math. Comput. **19**(90), 297–301, 1965.

[88] M.E. Marhic, "Discrete Fourier transforms by single-mode star networks," Opt. Lett. **12**(1), 63–65, 1987.

[89] Y.-K. Huang, D. Qian, R. E. Saperstein, P.N. Ji, N. Cvijetic, L. Xu, and T. Wang, "Dual-polarization 2×2 IFFT/FFT optical signal processing for 100-Gb/s QPSK-PDM all-optical OFDM," in Proc. of Optical Fiber Commun. Conf., paper OTuM4, 2009.

[90] A.D. Ellis, and F.C.G. Gunning, "Spectral density enhancement using coherent WDM," IEEE Photon. Technol. Lett., **17**(2), 504–506, 2005.

[91] T. Healy, F.C.G. Gunning, A.D. Ellis, and J.D. Bull, "Multi-wavelength source using low drive-voltage amplitude modulators for optical communications," Opt. Express **15**(6), 2981–2986, 2007.

[92] S.J.B. Yoo, "Wavelength conversion technologies for WDM network applications," J. Lightw. Technol. **14**(6), 955–966, 1996.

[93] Y. Liu, E. Tangdiongga, Z. Li, H. de Waardt, A.M.J. Koonen, G.D. Khoe, X. Shu, I. Bennion, and H.J.S. Dorren, "Error-free 320Gb/s all-optical wavelength conversion using a single semiconductor optical amplifier," J. Lightw. Technol. **25**(1), 103–108, 2007.

[94] M.L. Nielsen, B. Lavigne, and B. Dagens, "Polarity-preserving SOA-based wavelength conversion at 40 Gb/s using bandpass filtering," Electron. Lett. **39**(18), 1334–1335, 2003.

[95] J. Leuthold, B. Mikkelsen, G. Raybon, C.H. Joyner, J.L. Pleumeekers, B.I. Miller, K. Dreyer, and R. Behringer, "All-optical wavelength conversion between 10 and

100 Gb/s with SOA delayed-interference configuration," Opt. Quantum Electron. **33**(7-10), 939–952, 2001.

[96]   S. Nakamura, Y. Ueno, and K. Tajima, "168-Gb/s all-optical wavelength conversion with a symmetric-Mach-Zehnder-type switch," IEEE Photon. Technol. Lett. **13**(10), 1091–1093, 2001.

[97]   J. Leuthold, D.M. Marom, S. Cabot, J.J. Jaques, R. Ryf, and C.R. Giles, "All-optical wavelength conversion using a pulse reformatting optical filter," J. Lightw. Technol. **22**(1), 186–192, 2004.

[98]   A. V. Uskov, J. Mørk, B. Tromborg, T.W. Berg, I. Magnusdottir, and E.P. O'Reilly, "On high-speed cross gain modulation without pattern effects in quantum dot semiconductor optical amplifiers," Opt. Commun. **227**(4-6), 363–369, 2003.

[99]   P. Borri, W. Langbein, J.M. Hvam, F. Heinrichsdorf, M.-H. Mao, and D. Bimberg, "Ultrafast gain dynamics in InAs-InGaAs quantum dot amplifiers," IEEE Photon. Technol. Lett. **12**(6), 594–596, 2000.

[100] M. Sugawara, T. Akiyama, N. Hatori, Y. Nakata, H. Ebe, and H. Ishikawa, "Quantum-dot semiconductor optical amplifiers for high-bit-rate signal processing up to 160 Gb/s and a new scheme of 3R regenerators," Meas. Sci. Technol. **13**(11), 1683–1691, 2002.

[101] T. Vallaitis, C. Koos, R. Bonk, W. Freude, M. Laemmlin, C. Meuer, D. Bimberg, and J. Leuthold, "Slow and fast dynamics of gain and phase in a quantum dot semiconductor optical amplifier," Opt. Express **16**(1), 170–178, 2008.

[102] T. Akiyama, M. Sugawara, and Y. Arakawa, "Quantum-dot semiconductor optical amplifiers," in Proc. of IEEE **95**(9), 1757–1766, 2007.

[103] R. Bonk, P. Vorreau, S. Sygletos, T. Vallaitis, J. Wang, W. Freude, and J. Leuthold, "An interferometric configuration for performing cross-gain modulation with improved signal quality," in Proc. of Optical Fiber Commun. Conf., paper JWA70, 2008.

[104] A.V. Oppenheim, and R.W. Schafter, "Digital signal processing," Englewood Cliffs, NJ: Prentice-Hall, 1975.

[105] C.K. Madsen, and J.H. Zhao, "Optical filter design and analysis: a signal processing approach," Wiley, New York, 1999.

[106] S. Betti, F. Curti, G. De Marchis, and E. Iannone, "A novel multilevel coherent optical system: four quadrature signaling," J. Lightw. Technol. **9**(4), 514–523, 1991.

[107] Y. Han, and G. Li, "Coherent optical communication using polarization multiple-input-multiple-output," Opt. Express **13**(19), 7527–7534, 2005.

Acknowledgements

At this point, I would like to acknowledge those who offered me their valuable support and encouragement during the work of this thesis.

First of all, I want to thank my supervisors Prof. Juerg Leuthold and Prof. Wolfgang Freude for providing guidance and funding for my work. Prof. Leuthold I thank for his broad vision and enthusiasm in my work. His firm confidence on my work is one of the main sources sustains me to face all the challenges and inspires me to find new methods. Prof. Freude I thank for his spirit of precision on both science and language. His strict requirement on the manuscripts, his broaden knowledge and patience helps me to elaborate this work in a scientific and intelligible way. I also want to thank both of them for the time and large effort in working on our manuscripts. I would also like to thank Prof. Christian Koos for the discussion and his suggestions which lead to a success of this thesis.

I would like to thank Prof. Uli Lemmer who accepted me as a KSOP Ph.D. student and offer the support as a co-referee for this thesis. I thank Dr. Timo Mappes for his tutorship during my Ph.D. study. I also like to thank Dr. Judith Elsner, Vanessa Mayer, Eva Hildenbrand and all the secretaries of KSOP for their support in the administration of KSOP.

I would like to thank Dr. Ruediger Maestle of Agilent Technology (now with FISBA OPTIK AG) for the cooperation, discussion and quality control during the Agilent university relation project.

I would like to thank Dr. Sven Schuele, Stefan Hengsbach, Uwe Hollenback, and Dr. Jungen Mohr of IMT KIT for their support on the LIGA bench fabrication and building the alignment setup for the micro-optical interferometer. Dr. Sven Schuele I also thank for his cooperation in filing the patent and co-supervision of a student thesis. Stefan I thank for his assistance in the optical lab at IMT.

I would like to thank Dr. Shalva Ben-Ezra of Finisar Corporation, Israel, for his consistence support on my work with the indispensable devices and equipment.

I would like to thank Dr. Carsten Schmidt-Langhorst of Fraunhofer Institute for Telecommunications, Heinrich-Hertz-Institut (HHI) for his collaboration during Euro-fos joint experiment activities. The discussions and suggestions have brought significant contribution to my work. I thank him for sharing his knowledge and the measurement equipment during the demonstration of self-coherent receiver algorithms.

I would like to thank Prof. Moshe Nazarathy of Technion, Israel Institute of Technology, for his pioneering research and the discussion and exchange of ideas on self-coherent receiver. I thank for his analytical thinking and fast responding to all my emails.

Dr. Sami Mumtaz, and Prof. Yves Jaouen of Telecom Paris-Tech I thank for their collaboration during a Euro-fos joint experiment activity on novel coding for OFDM signals.

Dr. Bernd Nebendahl of Agilent Technology I thank for the optical modulation analyzer he loaned to us during the experiments and his advice on the use of the instrument.

Prof. Hercules Avramopoulos, Dr. Christos Kouloumentas, and Dr. Costis Christogiannis of Nation Technical University of Athens, Greece, I thank for their work in administration of the Euro-fos project. Dr. Colja Schubert of HHI, Dr. Jorge Seoane of DTU, and Dr. Didier Erasme of Telecom Paris-Tech, I thank for their work in coordination of the work packages in Euro-fos project.

The mechanical workshop of IPQ, Hans Bürger, Manfred Hirsch, Werner Höhne, and the apprentices I thank for their great support for the construction of the Kovar bench for the delay interferometers and all the necessary mechanical structures for holding and adjusting the optical elements. Oswald Speck of the packaging lab I thank for the great support in optical elements glue, fiber splicing, and assistance in the optical packaging lab. Werner Podszus, and Johann Hartwig Hauschild, I thank for their support in building the power supply for the delay interferometers. Martin Winkeler and Sebastian Struck I thank for their support in maintaining the computer system and networks.

My former colleagues have helped me a lot with their knowledge and experience. I thank Dr. Jin Wang, Andrej Marculescu, who taught me the lab work and simulation tools and provided me great support during my Master thesis. Dr. Philipp Vorreau I thank for his support in the system lab and his knowledge in the RF electronics. Dr. Stelios Sygletos, Dr. Thomas Vallaitis, and Dr. Rene Bonk I thank for the collaboration in the all optical wavelength converter projects. Dr. Marcus Winter I thank for his support in Euro-fos project and his knowledge in OFDM systems. Dr. Ayan Maitra I thank for the discussion during my Master thesis. Dr. Shunfeng Li I thank for his friendship and hospitality of his family. Dr. Jan-Michael Brosi I thank for introducing me the simulation tools for integrated optics. Dr. Gunnar Böttger, Dr. Arvind Mishra, and Martin Moch I thank for the interesting discussions and their kindness to share the ideas of their work with me.

For the great collaboration I want to thank all my colleagues at IPQ. David Hillerkuss I thank for his support of all system-level experiments. René Schmogrow I thank his indispensable software defined transmitter and his support during the experiments. Swen Koenig, Moritz Röger, Christos Klamouris, and Claudius Weimann it was a great fun to stay in the same office with you. Swen Koenig I thank for his support in using the arbitrary waveform generator and optical modulation analyzer. Alexandra Ludwig I thank for her help in assembling the balanced receiver and optical hybrid. Philipp Schindler I thank his support in the experiments, co-supervision of a student project, and help me in translation of the abstract of this thesis in German. Luca Alloatti, Nicole Lindenmann, Rober Palmer and Simon Schneider I thank for their support in tutoring the OKT student lab. Dr. Sean O'Duill and Philipp Schindler I thank for their help in correcting the wording and grammar of this thesis. Dietmar Korn, Argishti Melikyan, Sascha Muehlbrandt, Djorn Karnick, Dr. Sebastian Koeber, and Joerg Pfeifle I thank for the kind atmosphere we share in the past 5 years.

Many thanks to all my students who have contributed to our work: Andrew Efremov, Zhenhao Zhang, Kai Worms, Benjamin Rechiter, Matthias Lauermann, Hammam Shakhtour, Binbin Li, Amandev Singh, Qiang Shi, Zhao Wang, Ahmad Mustafa, Wahab Shah, Muhammad Rodlin Billah, Soumya Sunder Dash, and Lu Zhou.

Many thanks to our secretaries, Bernadette Lehmann, Ilse Kober, Andrea Riemensperger, and Angelika Olbrich for their caring personalities and the help in administrative matters.

Besides the work, I also would like to thank to all the soccer players in IPQ and IHE, especially the organizer Moritz Röger. It's a lot fun to play with you. I would also like to thank to all my friends in Karlsruhe, and the Chinese community here. It is because of you I feel less homeless.

Last but not least, I thank my family, my parents Dr. Weizhong Li and Yuwei Sui. In the last 5 years we were departed by over 10000 km but your care is always around. I also would like to thank my dear Cheng for all the time we spent together and your affection.

This work was supported by the Agilent University Relation Program, the European Commission's Network of Excellence *EuroFOS* and Karlsruhe School of Optics and Photonics (KSOP). This work was further supported in part by the OTONES trans-national Piano+ EU program and by the Israeli Science Foundation (ISF). I further acknowledge the Open Access Publishing Fund of the Karlsruhe Institute of Technology (KIT).

List of Publications

Patent

[P1]   **J. Li**, M. Lauermann, S. Schuele, J. Leuthold, and W. Freude, "Optical detector for detecting optical signal beams, method to detect optical signals, and use of an optical detector to detect optical signals," US patent 20, 120, 224, 184, 2012.

Journal Papers

[J1]   **J. Li**, M. R. Billah, P. C. Schindler, M. Lauermann, S. Schuele, S. Hengsbach, U. Hollenbach, J. Mohr, C. Koos, W. Freude, and J. Leuthold, Opt. Express, submitted.

[J2]   **J. Li**, R. Schmogrow, D. Hillerkuss, P. Schindler, M. Nazarathy, C. Schmidt-Langhorst, S.-B. Ezra, C. Koos, W. Freude, and J. Leuthold, "A self-coherent receiver for PolMUX-signals," Opt. Express, vol. 20, no. 19, pp. 21413-21433, 2012.

[J3]   I. Tselniker, M. Nazarathy, S.-B. Ezra, **J. Li**, and J. Leuthold, "Self-coherent complex field reconstruction with in-phase and quadrature delay detection without a direct-detection branch," Opt. Express, vol. 20, no. 14, pp. 15452–15473, 2012.

[J4]   D. Hillerkuss, R. Schmogrow, T. Schellinger, M. Jordan, M. Winter, G. Huber, T. Vallaitis, R. Bonk, P. Kleinow, F. Frey, M. Roeger, S. Koenig, A. Ludwig, A. Marculescu, **J. Li**, M. Hoh, M. Dreschmann, J. Meyer, S. Ben Ezra, N. Narkiss, B. Nebendahl, F. Parmigiani, P. Petropoulos, B. Resan, A. Oehler, K. Weingarten, T. Ellermeyer, J. Lutz, M. Moeller, M. Huebner, J. Becker, C. Koos, W. Freude, and J. Leuthold, "26 Tbit s−1 line-rate super-channel transmission utilizing all-optical fast Fourier transform processing," Nature Photonics, vol 5, pp. 364–371, 2011.

[J5]   **J. Li**, K. Worms, R. Maestle, D. Hillerkuss, W. Freude, and J. Leuthold, "Free-space optical delay interferometer with tunable delay and phase," Opt. Express, vol. 19, no. 12, pp. 11654-11666, 2011.

[J6]   L. Alloatti, D. Korn, R. Palmer, D. Hillerkuss, **J. Li**, A. Barklund, R. Dinu, J. Wieland, M. Fournier, J. Fedeli, H. Yu, W. Bogaerts, P. Dumon, R. Baets, C. Koos, W. Freude, and J. Leuthold, "42.7 Gbit/s electro-optic modulator in silicon technology," Opt. Express, vol. 19, no. 12, pp. 11841–11851, 2011.

[J7]   D. Hillerkuss, M. Winter, M. Teschke, A. Marculescu, **J. Li**, G. Sigurdsson, K. Worms, S. Ben Ezra, N. Narkiss, W. Freude, and J. Leuthold, "Simple all-optical FFT scheme enabling Tbit/s real-time signal processing," Opt. Express, vol. 18, no. 9, pp. 9324–9340, 2010.

[J8]   T. Vallaitis, R. Bonk, J. Guetlein, D. Hillerkuss, **J. Li**, R. Brenot, F. Lelarge, G. H. Duan, W. Freude, and J. Leuthold, "Quantum dot SOA input power dynamic range improvement for differential-phase encoded signals," Opt. Express, vol. 18, no. 6, pp. 6270–6276, 2010.

[J9]   S. Sygletos, R. Bonk, T. Vallaitis, A. Marculescu, P. Vorreau, **J. Li**, R. Brenot, F. Lelarge, H. Duan, W. Freude, and J. Leuthold, "Filter assisted wavelength conversion with quantum-dot SOAs," J. Lightw. Technol., vol. 28, no. 6, pp. 882–897, 2010.

[J10] J. Wang, A. Marculescu, **J. Li**, P. Vorreau, S. Tzadok, S. Ben Ezra, S. Tsadka, W. Freude, and J. Leuthold, "Pattern effect removal technique for semiconductor optical amplifier-based wavelength conversion," IEEE Photon. Technol. Lett., vol. 19, no. 24, pp. 1955–1957, 2007.

Conference Contributions

[C1]   W. Freude, L. Alloatti, A. Melikyan, R. Palmer, D. Korn, N. Lindenmann, T. Vallaitis, D. Hillerkuss, **J. Li**, A. Barklund, R. Dinu, J. Wieland, M. Fournier, J. Fedeli, S. Walheim, P. M. Leufke, S. Ulrich, J. Ye, P. Vincze, H. Hahn, H. Yu, W. Bogaerts, P. Dumont, R. Baets, B. Breiten, F. Diederich, M. T. Beels, I. Biaggio, Th. Schimmel, C. Koos, and J. Leuthold, "Nonlinear optics on the silicon platform," in Proc. of Optical Fiber Communication Conference (OFC'12), Los Angeles (CA), USA, paper OTh3H.6, 2012 [invited].

[C2]   C. Koos, L. Alloatti, D. Korn, R. Palmer, T. Vallaitis, R. Bonk, D. Hillerkuss, **J. Li**, W. Bogaerts, P. Dumon, R. Baetes, M. L. Scimeca, I. Biaggio, A. Barklund, R. Dinu, J. Wieland, M. Fournier, J. Fedeli, W. Freude, and J. Leuthold, "Silicon nanophotonics and silicon-organic hybrid (SOH) integration," in Proc. of General Assembly and Scientific Symposium, 2011.

[C3]   C. Koos, L. Alloatti, D. Korn, R. Palmer, D. Hillerkuss, **J. Li**, A. Barklund, R. Dinu, J. Wieland, M. Fournier, J. Fedeli, H. Yu, W. Bogaerts, P. Dumon, R. Baets, W. Freude, and J. Leuthold, "Silicon-organic hybrid (SOH) electro-optical devices," in Proc. of Integrated Photonics Research, Silicon and Nano Photonics (IPR) Topical Meeting of the OSA, Toronto, Canada, (Optical Society of America), paper IWF1, 2011 [invited].

[C4]   S. Mumtaz, G. Rekaya-Ben Othman, Y. Jaouen, **J. Li**, S. Koenig, R. Schmogrow, and J. Leuthold, "Alamouti code against PDL in polarization multiplexed systems," in Proc. of Signal Processing in Photonic Communications, OSA Technical Digest (Optical Society of America, 2011), paper SPTuA2, 2011.

[C5]   S. Mumtaz, **J. Li**, S. Koenig, Y. Jaouen, R. Schmogrow, G. Rekaya-Ben Othman, and J. Leuthold, "Experimental demonstration of PDL mitigation using polarization-time coding in PDM-OFDM systems," in Proc. of Signal Processing in Photonic Communications, OSA Technical Digest (Optical Society of America, 2011), paper SPWB6, 2011.

[C6] **J. Li**, C. Schmidt-Langhorst, R. Schmogrow, D. Hillerkuss, M. Lauermann, M. Winter, K. Worms, C. Schubert, C. Koos, W. Freude, and J. Leuthold, "Self-coherent receiver for PolMUX coherent signals," in Proc. of Optical Fiber Communication Conference (OFC'11), Los Angeles (CA), USA, paper OWV5, 2011.

[C7] W. Freude, D. Hillerkuss, T. Schellinger, R. Schmogrow, M. Winter, T. Vallaitis, R. Bonk, A. Marculescu, **J. Li**, M. Dreschmann, J. Meyer, S. Ben Ezra, M. Caspi, B. Nebendahl, F. Parmigiani, P. Petropoulos, B. Resan, A. Oehler, K. Weingarten, T. Ellermeyer, J. Lutz, M. Moeller, M. Huebner, J. Becker, C. Koos, and J. Leuthold, "Orthogonal frequency division multiplexing (OFDM) in photonic communications," in Proc. of 10th Intern. Conf. on Fiber Optics & Photonics (Photonics'08), Indian Institute of Technology Guwahati (IIT Guwahati), Guwahati, Assam, India, 2010 [invited].

[C8] W. Freude, L. Alloatti, T. Vallaitis, D. Korn, D. Hillerkuss, R. Bonk, R. Palmer, **J. Li**, T. Schellinger, M. Fournier, J. Fedeli, W. Bogaerts, P. Dumon, R. Baets, A. Barklund, R. Dinu, J. Wieland, and M. Scimeca, "High-speed signal processing with silicon-organic hybrid devices," Proc. of European Optical Society Annual Meeting (EOS'10), Parc Floral De Paris, France, 2010 [invited].

[C9] L. Alloatti, D. Korn, D. Hillerkuss, T. Vallaitis, **J. Li**, R. Bonk, R. Palmer, T. Schellinger, A. Barklund, R. Dinu, J. Wieland, M. Fournier, J. Fedeli, W. Bogaerts, P. Dumon, R. Baets, C. Koos, W. Freude, and J. Leuthold, "Silicon high-speed electro-optic modulator," in Proc. of Group IV Photonics 2010, China, paper ThC2, 2010.

[C10] L. Alloatti, D. Korn, D. Hillerkuss, T. Vallaitis, **J. Li**, R. Bonk, R. Palmer, T. Schellinger, A. Barklund, R. Dinu, J. Wieland, M. Fournier, J. Fedeli, P. Dumon, R. Baets, C. Koos, W. Freude, and J. Leuthold, "40 Gbit/s Silicon-organic hybrid (SOH) phase modulator," in Proc. of European Conference on Optical Communication (ECOC'10), Torino, Italy, paper Tu.5.C.4, 2010.

[C11] J. Leuthold, M. Winter, W. Freude, C. Koos, D. Hillerkuss, T. Schellinger, R. Schmogrow, T. Vallaitis, R. Bonk, A. Marculescu, **J. Li**, M. Dreschmann, J. Meyer, M. Huebner, J. Becker, S. Ben Ezra, N. Narkiss, B. Nebendahl, F. Parmigiani, P. Petropoulos, B. Resan, A. Oehler, K. Weingarten, T. Ellermeyer, J. Lutz, and M. Moeller, "All-optical FTT signal processing of a 10.8 Tb/s single channel OFDM signal," in Proc. of Photonics in Switching (PS), OSA Technical Digest (Optical Society of America), paper PWC1, 2010 [invited].

[C12] J. Leuthold, D. Hillerkuss, M. Winter, **J. Li**, K. Worms, C. Koos, W. Freude, S. Ben Ezra, and N. Narkiss, "Terabit/s FFT Processing – Optics can do it on-the Fly," in Proc. of 12th Intern. Conf. on Transparent Optical Networks (ICTON'10), München, Germany, paper Mo.D1.4, 2010 [invited].

[C13] **J. Li**, K. Worms, A. Marculescu, D. Hillerkuss, S. Ben-Ezra, W. Freude, and J. Leuthold, "Optical vector signal analyzer based on differential detection with inphase and quadrature phase control," in Proc. of OSA Optics & Photonics Congress, Karlsruhe, Germany, paper SPTuB3, 2010.

[C14] D. Hillerkuss, M. Winter, M. Teschke, A. Marculescu, **J. Li**, G. Sigurdsson, K. Worms, W. Freude, and J. Leuthold, "Low-complexity optical FFT scheme enabling Tbit/s all-optical OFDM communication," in Proc. of 11. VDE-ITG-Fachtagung Photonische Netze, Leipzig, 2010.

[C15] S. Schuele, S. Hengsbach, U. Hollenbach, **J. Li**, J. Leuthold, and J. Mohr, "Active modular microsystems based on Mach-Zehnder interferometers," in Proc. of SPIE Photonics Europe, Nonlinear Optics and its Applications, vol. 7716, paper 7716–36, 2010.

[C16] D. Hillerkuss, A. Marculescu, **J. Li**, M. Teschke, G. Sigurdsson, K. Worms, S. Ben Ezra, N. Narkiss, W. Freude, and J. Leuthold, "Novel optical fast Fourier transform scheme enabling real-time OFDM processing at 392 Gbit/s and beyond," in Proc. of Optical Fiber Communication Conference (OFC'10), San Diego (CA), USA, paper OWW3, 2010 [Best Student Paper Award].

[C17] D. Hillerkuss, T. Schellinger, R. Schmogrow, M. Winter, T. Vallaitis, R. Bonk, A. Marculescu, **J. Li**, M. Dreschmann, J. Meyer, S. Ben Ezra, N. Narkiss, B. Nebendahl, F. Parmigiani, P. Petropoulos, B. Resan, K. Weingarten, T. Ellermeyer, J. Lutz, M. Moeller, M. Hübner, J. Becker, C. Koos, W. Freude, and J. Leuthold, "Single source optical OFDM transmitter and optical FFT receiver demonstrated at line rates of 5.4 and 10.8 Tbit/s," in Proc. of Optical Fiber Communication Conference (OFC'10), San Diego (CA), USA, postdeadline paper PDPC1, 2010.

[C18] R. Bonk, T. Vallaitis, J. Guetlein, D. Hillerkuss, **J. Li**, W. Freude, and J. Leuthold, "Quantum dot SOA dynamic range improvement for phase modulated signals," in Proc. of Optical Fiber Communication Conference (OFC'10), San Diego (CA), USA, paper OThK3, 2010.

[C19] **J. Li**, K. Worms, D. Hillerkuss, B. Richter, R. Maestle, W. Freude, and J. Leuthold, "Tunable free space optical delay interferometer for demodulation of differential phase shift keying signals," in Proc. of Optical Fiber Communication Conference (OFC'10), San Diego (CA), USA, paper JWA24, 2010.

[C20] T. Vallaitis, R. Bonk, J. Guetlein, D. Hillerkuss, **J. Li**, W. Freude, J. Leuthold, C. Koos, M. L. Scimeca, I. Biaggio, F. Diederich, B. Breiten, P. Dumon, and R. Baets, "All-optical wavelength conversion of 56 Gbit/s NRZ-DQPSK signals in silicon-organic hybrid strip waveguides," in Proc. of Optical Fiber Communication Conference (OFC'10), San Diego (CA), USA, paper OTuN1, 2010.

[C21] **J. Li**, K. Worms, P. Vorreau, D. Hillerkuss, A. Ludwig, R. Maestle, S. Schüle, U. Hollenbach, J. Mohr, W. Freude, and J. Leuthold, "Optical vector signal analyzer based on differential direct detection," in Proc.of 22nd Annual Meeting of the IEEE Photonics Society (LEOS 2009), Belek-Antalya (Turkey), paper TuA4, 2009 (Best Student Paper Award).

[C22] T. Vallaitis, D. Hillerkuss, **J.-S. Li**, R. Bonk, N. Lindenmann, P. Dumon, R. Baets, M. L. Scimeca, I. Biaggio, F. Diederich, C. Koos, W. Freude, and J. Leuthold, "All-optical wavelength conversion using cross-phase modulation at 42.7 Gbit/s in silicon-organic hybrid (SOH) waveguides", in Proc. of Intern. Conf. on Photonics in Switching (PS'09), Pisa, Italy, postdeadline paper PDP3, 2009.

[C23] S. Schüle, S. Hengsbach, U. Hollenbach, **J. Li**, J. Leuthold, S. Schonhardt, U. Wallrabe, and J. Mohr, "Adaptive micro-optical interferometer for different measurement and network applications," in Proc. of OEPT 2009 Conference, Orlando, Florida, USA, pp. 287–291, 2009.

[C24] S. Sygletos, R. Bonk, T. Vallaitis, A. Marculescu, P. Vorreau, **J. Li**, R. Brenot, F. Lelarge, G. H. Duan, W. Freude, and J. Leuthold, "Optimum filtering schemes for performing wavelength conversion with QD-SOA," in Proc. of 11th Intern. Conf. on Transparent Optical Networks (ICTON'09), Ponta Delgada, Island of Sao Miguel, Portugal, vol. 1, pp. 1–4, paper Mo.C1.3, 2009 [invited].

[C25] R. Bonk, S. Sygletos, R. Brenot, T. Vallaitis, A. Marculescu, P. Vorreau, **J. Li**, W. Freude, F. Lelarge, G.-H. Duan, and J. Leuthold, "Optimum filter for wavelength conversion with QD-SOA," in Proc. of CLEO/IQEC, Baltimore (Maryland), USA, paper CMC6, 2009.

[C26] A. Marculescu, S. Sygletos, **J. Li**, D. Karki, D. Hillerkuss, S. Ben-Ezra, S. Tsadka, W. Freude, and J. Leuthold, "RZ to CSRZ format and wavelength conversion with regenerative properties," in Proc. of Optical Fiber Communication Conference (OFC'09), San Diego (CA), USA, paper OThS1, 2009.

[C27] **J. Li**, J. Wang, A. Marculescu, P. Vorreau, Z. Zhang, W. Freude, and J. Leuthold, "All-Optical SOA-based wavelength converter assisted by optical filters with wide operation wavelength and large dynamic input power range," in Proc. of Asia-Pacific Optical Communications 2008 (APOC'08), Hangzhou, China, paper 7135-82, 2008.

[C28] J. Wang, A. Marculescu, **J. Li**, Z. Zhang, W. Freude, and J. Leuthold, "All-optical vestigial-sideband signal generation and pattern effect mitigation with an SOA based red-shift optical filter wavelength converter," in Proc. of European Conference on Optical Communication, (ECOC'08), Brussels, paper We.2.C.6, 2008.

[C29] S. Schüle, U. Hollenbach, **J. Li**, J. Leuthold, and J. Mohr, "Set up of a modular micro-optical system based on free space optics," in Proc. of MOC 2008 Conference, Bruxelles, Belgium, pp.334–335, 2008.

[C30] S. Schüle, U. Hollenbach, J. Mohr, **J. Li**, P. Vorreau, A. Efremov, J. Leuthold, and S. Schonhardt, "Modular integration of microactuators and micro-optical benches," in Proc. of Conference on Micro-Optics Strasbourg, FRANCE, 2008 [invited].

[C31] A. Marculescu, J. Wang, **J. Li**, P. Vorreau, S. Zadok, S. Ben Ezra, S. Tsadka, W. Freude, and J. Leuthold, "Pattern effect removal technique for semiconductor optical amplifier-based wavelength conversion," in Proc. of European Conference on Optical Communication (ECOC'07), Berlin, Germany, paper Tu3.4.6, 2007.

Curriculum Vitae

Jingshi Li
born on November 8th, 1983
in Liaoning, P. R. China
Citizenship: Chinese

jingshili@gmail.com

Education

08/2007–11/2012	**University of Karlsruhe (TH) / Karlsruhe Institute of Technology (KIT), Institute of Photonics and Quantum Electronics (IPQ)**
	Ph.D. thesis on "Optical Delay Interferometers and their Application for Self-coherent Detection"
	Date of defense: Nov. 12th 2012.
	Grade: 1.0 mit Auszeichnung (Summa Cum Laude)
08/2005–05/2007	**KTH - Royal Institute of Technology, Sweden**
	Photonics major, Master of Science in Electrical Engineering
	Grade: 4.43 / 5.00
	Final thesis: "All-Optical Wavelength Conversion based on a single Semiconductor Optical Amplifier assisted by a Pulse Reformatting Optical Filter"
	Grade: Very good
09/2001–06/2005	**Zhejiang University, P. R. China**
	Optical Engineering major, Bachelor of Science
	Grade: 3.36 / 4.00
	Final thesis: "Design and Implementation of LED Displays"
	Grade: Excellent
12/1998–07/2001	**Dalian Yuming Senior High School, P. R. China**
09/1998–12/1998	**Senior High School Affiliated to Dalian University of Technology, P. R. China**
09/1995–07/1998	**Middle School Affiliated to Dalian University of Technology, P. R. China**

Karlsruhe Series in Photonics & Communications
KIT, Institute of Photonics and Quantum Electronics (IPQ)
(ISSN 1865-1100)

Die Bände sind unter www.ksp.kit.edu als PDF frei verfügbar
oder als Druckausgabe bestellbar.

Band 1   Christian Koos
**Nanophotonic Devices for Linear and Nonlinear Optical
Signal Processing.** 2008
ISBN 978-3-86644-178-1

Band 2   Ayan Maitra
Nonlinear Resonators for All-Optical Signal Processing. 2009
ISBN 978-3-86644-150-7

Band 3   Jin Wang
**Pattern Effect Mitigation Techniques for All-Optical Wavelength
Converters Based on Semiconductor Optical Amplifiers.** 2008
ISBN 978-3-86644-276-4

Band 4   Jan-Michael Brosi
**Slow-Light Photonic Crystal Devices for High-Speed Optical
Signal Processing.** 2009
ISBN 978-3-86644-313-6

Band 5   Christoph Dyroff
**Tunable Diode-Laser Absorption Spectroscopy for Trace-Gas
Measurements with High Sensitivity and Low Drift.** 2009
ISBN 978-3-86644-328-0

Band 6   Philipp Vorreau
**An Optical Grooming Switch for High-Speed Traffic
Aggregation in Time, Space and Wavelength.** 2010
ISBN 978-3-86644-502-4

Band 7   Thomas Vallaitis
**Ultrafast Nonlinear Silicon Waveguides and Quantum
Dot Semiconductor Optical Amplifiers.** 2011
ISBN 978-3-86644-748-6

Band 8   René Bonk
**Linear and Nonlinear Semiconductor Optical Amplifiers for
Next-Generation Optical Networks.** 2013
ISBN 978-3-86644-956-5

**Karlsruhe Series in Photonics & Communications
KIT, Institute of Photonics and Quantum Electronics (IPQ)
(ISSN 1865-1100)**

Band 9   David Hillerkuss
           Single-Laser Multi-Terabit/s Systems. 2013
           ISBN 978-3-86644-991-6

Band 10   Moritz Röger
           Optically Powered Highly Energy-efficient Sensor Networks. 2013
           ISBN 978-3-86644-972-5

Band 11   Jingshi Li
           **Optical Delay Interferometers and their Application
           for Self-coherent Detection.** 2013
           ISBN 978-3-86644-957-2